Chemistry

An Introduction for
Medical and Health Sciences

Alan Jones

*Formerly Head of Chemistry and Physics
Nottingham Trent University*

John Wiley & Sons, Ltd

Other Wiley Editorial Offices

John Wiley & Sons Inc., 111 River Street, Hoboken, NJ 07030, USA

Jossey-Bass, 989 Market Street, San Francisco, CA 94103-1741, USA

Wiley-VCH Verlag GmbH, Boschstr. 12, D-69469 Weinheim, Germany

John Wiley & Sons Australia Ltd, 33 Park Road, Milton, Queensland 4064, Australia

John Wiley & Sons (Asia) Pte Ltd, 2 Clementi Loop #02-01, Jin Xing Distripark, Singapore 129809

John Wiley & Sons Canada Ltd, 22 Worcester Road, Etobicoke, Ontario, Canada M9W 1L1

Wiley also publishes its books in a variety of electronic formats. Some content that appears in print may not
be available in electronic books.

Library of Congress Cataloging-in-Publication Data

Jones, Alan, 1941–
    Chemistry : an introduction for medical and health sciences / Alan Jones
    p. cm.
Includes bibliographical references and index.
ISBN   0-470-09288-2 (cloth) – 0-470-09289-0

1. Biochemistry. 2. Chemistry. 3. Pharmaceutical chemistry. II. Title
    QP514.2.J66   2005
    612′.015–dc22   2004029124

British Library Cataloguing in Publication Data

A catalogue record for this book is available from the British Library

ISBN 0 470 09288 2 hardback
ISBN 0 470 09289 0 paperback

Typeset in 11/14pt Times by Thomson Press (India) Limited, New Delhi
Printed and bound in Great Britain by Antony Rowe Ltd, Chippenham, Wiltshire
This book is printed on acid-free paper responsibly manufactured from sustainable forestry
in which at least two trees are planted for each one used for paper production.

Chemistry

An Introduction for
Medical and Health Sciences

Contents

Preface

Recent years have seen significant changes in the practice, education and training of doctors, medical, nursing and healthcare professionals. Pieces of paper are required to show competency in a wide range of skills. There is also a requirement for continuing professional development in order that people increase their knowledge and skills. The United Kingdom Central Council for Nursing, Midwifery and Health Visiting publication *Fitness for Practice* notes that there will be: 'greater demands upon nurses and midwives for technical competence and scientific rationality'.

The daily use of chemicals in the form of medicines and drugs means that there is a need for a basic understanding of chemistry. Do not be put off by this, as you will not be expected to be a chemical expert, but you will need to have some knowledge of the various chemicals in common medical use. You will not be expected to write complicated formulae or remember the structures of the drugs you administer, but it will be of use to know some of their parameters. Modern healthcare is becoming increasingly scientific, so there is a necessity to have a good introduction to chemical concepts. Scientific and chemical understanding leads to better informed doctors, nurses and healthcare workers.

This book starts each chapter with a self-test to check on chemical understanding, and then proceeds to move through the subject matter, always within the context of current practice. Anyone able to pass well on the self-test can move onto the next chapter. I hope you will find the Glossary a useful reference source for a number of chemical terms.

Finally, I would like to thank Mike Clemmet for his valuable contributions to earlier versions of the book, also Dr Sheelagh Campbell of the University of Portsmouth who reviewed the draft manuscript, and Malcolm Lawson-Paul for drawing the cartoons. Perhaps he has learned a little more about chemistry along the way!

Alan Jones

Introduction

This book is intended to introduce some of the basic chemistry for the medical and healthcare professions. The material is suitable for any such course or as a refresher for people returning to the profession. It is designed to give a basic introduction to chemical terms and concepts and will develop the relevant chemistry of drugs and medicines in common use in later chapters.

It can be used as a self-teaching book since it contains diagnostic questions at the beginning of each chapter together with the answers, at the end of the chapter.

It can also be used to supplement the chemistry done on any suitable course. It is not a compendium or list of current drugs and their contents. It is also suitable for people who have a limited chemical knowledge as it starts with the basic concepts at the start of each chapter.

How to use the book

Read Chapter 1. Just read it through quickly. Do not worry about total understanding at this stage. Use it as an introduction or refresher course for chemical terminology

Take in the 'feeling' of chemistry' – and begin to understand the basic principles. Think, but do not stop to follow up any cross-references yet. Just read it through. That will take about twenty minutes.

When you've read this section through once, and thought about it, read it through again, a few days later, but this time take it more slowly. If you are unclear about the chemical words used in Chapter 1 and the others Chapters, use the Glossary at the end of the book for clarification. After reading the whole of Chapter 1 you will be ready for a more detailed study of the relevant areas of chemistry in later chapters.

At the start of each chapter there are some diagnostic questions. If you get more than 80 % of the questions right (the answers are given at the end of each chapter),

Chemistry: An Introduction for Medical and Health Sciences, A. Jones
© 2005 John Wiley & Sons, Ltd

you probably understand the principles. Be honest with yourself. If you really feel that you do not understand it, talk to someone. Start with a fellow student. Then, if the two of you cannot sort it out, ask your lecturer/tutor – that is what they get paid for! You can always read the chapter again a little later. Sometimes familiarity with the words and concepts from a previous reading helps when you read it a second time. Remember this is a study book for your own professional development not a novel where it does not matter if you cannot remember the characters' names.

It will also be helpful, whenever needed or as an aid to your memory, to check on things by looking up words, concepts and definitions in the Glossary. Keep a notebook handy to jot down useful items to remember later.

Throughout the book, as you would expect, there are formulae and structures of chemical compounds. You need not remember these but they are included to show the principles being covered. You are not expected to work out the names of these compounds or balance equations but after a while some might stick in your memory.

In each of the later chapters there are 'scene setters' for the concepts covered in the chapters. The chapters start up with basic ideas and lead onto more detailed chemistry and applications.

Anyway, here we go! Enjoy it! I did when I wrote it and even later when I re-read it. Excuse my sense of humour; I feel it is needed when studying chemistry.

1 Starting Chemistry

Learning objectives

- To introduce some of the most relevant and commonly used chemical concepts, processes and naming systems.

- To show some of the background upon which medicinal chemistry is based.

Diagnostic test

Try this short test. If you score more than 80% you can use the chapter as a revision of your knowledge. If you score less than 80% you probably need to work through the text and test yourself again at the end using the same test. If you still score less than 80% then come back to the chapter after a few days and read it again.

1. What is the main natural source of drug material for research? (1)

2. What charge has each of the following particles: proton, neutron, electron? (3)

3. Covalent bonding gains its stability by what process? (1)

4. Ionic bonding gains its stability by what process? (1)

Chemistry: An Introduction for Medical and Health Sciences, A. Jones
© 2005 John Wiley & Sons, Ltd

5. From what natural source does aspirin originally come?   (1)

6. Who was the first person to come up with the idea of the atom? (1)

7. What is the arrangement called that puts all the elements into a logical pattern?   (1)

8. Who discovered penicillin?   (1)

Total 10 (80% = 8)
Answers at the end of the chapter.

1.1   Terminology and processes used in drug manufacture

The terms and nomenclature used in chemistry might seem over-complicated at first, but they have been internationally accepted. In this book we use the scientific names for chemicals, not their trivial or common names, e.g. ethanoic acid is used for acetic acid (a constituent of vinegar).

1.1.1   Separation and preparation of commonly used drugs

Where do drugs come from? Most people knows the story of the discovery of penicillin. In simplified form it tells that Alexander Fleming left a culture of bacteria in a Petri dish open in the laboratory. When he looked at it a few days later, he found a fungus or mould growing on it. There was a ring around each bit of the mould, where the bacteria had died. He decided that the mould must have produced a chemical that killed that bacteria. We might have said, 'Uch, dirty stuff' and thrown it out, but he realized he had discovered something new. He had discovered the first *antibiotic*. This all happened in the late 1920s, although it was not until the 1940s and World War II that it was used to great effect for treating infections.

The following section looks at some of the chemical principles which need to be considered when searching for a cure for a particular disease or condition. SARS in 2003 and the Bird Flu in Asia in 2004 were such examples where immediate new cures were sought to avoid a pandemic. The HIV virus has an uncanny knack of changing its surface proteins to confuse the drugs used in its treatments. Research is being conducted to overcome this problem.

As disease agents, such as MRSA, become more and more resistant to drugs, the search is on for new drugs to combat disease and attack viruses. Where should we look for new sources of combatants against disease? We should look where people have always looked – the natural drugs present in the plant world. There have always been 'witch doctors' and old women who have come up with concoctions which supposedly combat diseases, for example hanging garlic bags around a person's neck to drive away the plague, wearing copper bracelets to counteract arthritis or chewing the leaves of certain plants. Some of these remedies might have real significance.

Some of the most promising places to search for suitable plants are in the tropical rain forests, although even plants in places such as Milton Keynes seem to have medicinal uses, for example willow tree bark. The willow tree was the original source of aspirin-like medicines in Britain. It cured the pains from various complaints.

Herbal concoctions have been the basis of healing and also poisoning for centuries. Curare was used on the tips of poison darts to kill opponents, but in smaller quantities it was used as a muscle relaxant in surgery up to the 1960s.[1] Foxglove (digitalis) extracts, as well as being poisonous, have been found to help reduce blood pressure and aid people with heart problems. 'My mother-in-law used to wrap cabbage leaves around her arthritic knees to give her relief from pain just as her mother before had done'. In 2003 a short note in a British medical journal reported that this 'old wives tale' has been shown to have a scientific reason.[2]

Approximately 80 % of modern drugs came initially from natural sources. There are more different species of plants in the rain forests than in any other area on Earth. Many of these species are yet to be discovered and studied in detail. Every year, thousands of plant samples are collected by drug companies to find out whether they have any anti-disease activity. Many of them do. In the mean time, we continue to destroy the rain forests just to obtain teak furniture or some extra peanuts, but that is another story. This area of research is considered in more detail in Chapter 14.

The principles of how chemicals are isolated from plants will be used as an example. Aspirin has been chosen because it is one of the most widely used drugs in the world and it is also one of the most chemically simple, as well as one of the cheapest.

About 50 000 000 000 aspirin tablets are consumed each year throughout the world. On average, each adult takes the equivalent of 70 aspirin tablets (or tablets containing it) each year in the UK, but where did it all start?

Over 2400 years ago in ancient Greece, Hippocrates recommended the juice of willow leaves for the relief of pain in childbirth. In the first century AD in Greece, willow leaves were widely used for the relief of the pain of colic and gout. Writings from China, Africa and American Indians have all shown that they knew about the curative properties of the willow.

Try this one – it's good for the head after an all night session.

In 1763 the use of willow tree bark was reported in more specific terms by Reverend Edward Stone in a lecture to the Royal Society in London. He used its extracts to treat the fever resulting from malaria (then common in Britain; there are some marshes in the UK where the malarial mosquito still persists). He also found that it helped with 'the agues', probably what is now called arthritis. Other common medicines of the time included opium to relieve pain and Peruvian cinchona bark for fevers (it contained quinine).

In the early part of the 1800s chemists in Europe took willow leaves and boiled them with different solvents to try to extract the active ingredients. In 1825 an Italian chemist filtered such a solution and evaporated away the solvent. He obtained impure crystals of a compound containing some of the active ingredient. Repeated recrystallization and refinement of his experimental technique produced a pure sample of the unknown material (Figure 1.1).

In 1828 Buchner in Germany managed to obtain some pure white crystals of a compound by repeatedly removing impurities from an extract of willow bark. He called it 'salicin' (Figure 1.2). It had a bitter taste and relieved pain and inflammation. This same compound was extracted from a herb called meadowsweet by other chemists. Analysis of salicin showed it to be the active ingredient of willow bark joined to a sugar, glucose.

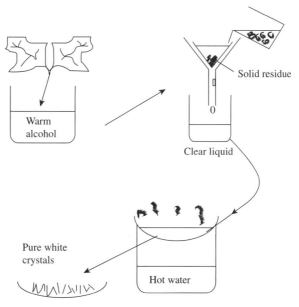

Figure 1.1 Separation of ingredients from willow

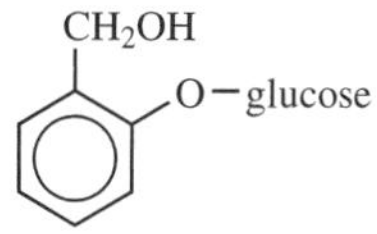

CH₂OH

O—glucose

Figure 1.2 Salicin

In the body, salicin is converted into salicylic acid (Figure 1.3) and it was this that was thought to be the active ingredient that relieved pain, but it had such a very bitter taste that it made some people sick. Some patients complained of severe irritation of the mouth, throat and stomach.

The extraction process for making the salicin also proved long and tedious and wasteful of trees: from 1.5 kg of willow bark only 30 g of salicin could be obtained.[3,4] Once the formula was known for salicylic acid, a group of chemists tried to work out how to make it artificially by a less expensive and tedious process.

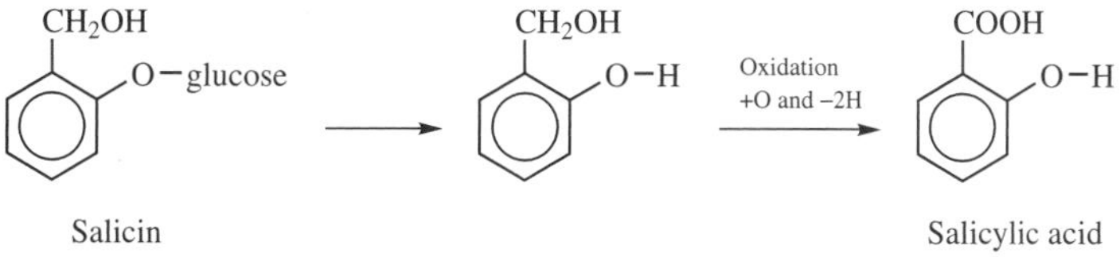

Salicin
Salicylic acid

Figure 1.3 Conversion of salicin to salicylic acid

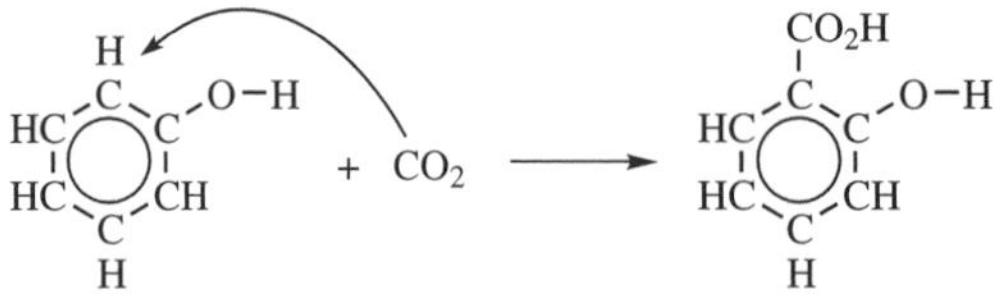

Figure 1.4 Synthesis of salicylic acid

It was not until 1860 that Professor Kolbe found a suitable way of doing this. He heated together phenol, carbon dioxide gas and sodium hydroxide (Figure 1.4; the hexagonal rings in the following figures are an abbreviation of a compound with carbon atoms joined to hydrogen atoms on each point of the hexagon). The phenol was extracted from coal tar and the carbon dioxide, CO_2, was readily made by heating limestone, a carbonate rock, or burning carbon:

$$CaCO_3 \rightarrow CaO + CO_2$$

Because of the ease of its synthesis it was beginning to look as though salicylic acid had a future as a pain-relieving drug, although it still had the drawback of its very bitter taste.

Felix Hoffman worked for the manufacturers Bayer. His father suffered from arthritis and became sick when he took salicylic acid. He challenged his son to find a better alternative. Hoffman did this in 1893 when he made the compound acetyl salicylic acid. This compound went through extensive clinical trials and in 1899 it came on the market as aspirin (Figure 1.5). It proved to be a wonder drug and still is.

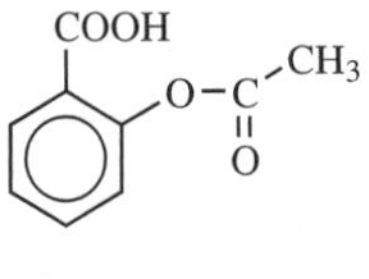

Figure 1.5 Aspirin

It is only in recent years that researchers have found out exactly how it works in our bodies. Previously all they knew was that it worked for a wide range of ailments, thinning the blood, lowering blood pressure and relieving pain for arthritis sufferers.

Aspirin deals with pain that comes from any form of inflammation, but it does cause some stomach bleeding. Therefore, research was undertaken for an alternative that was cheap to manufacture and would not cause stomach bleeding. This search led to the synthesis of paracetamol (Figure 1.6). Paracetamol does not cause stomach bleeding, but large doses damage the liver.

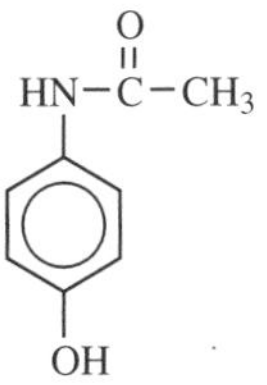

Figure 1.6 Paracetamol

A further series of drugs based on ibuprofen, developed by Boots in the 1980s, looks like being the most successful replacement for aspirin so far. Six hundred different molecules were made and tested before ibuprofen was perfected and clinically trialled. It is now sold over the counter and has few or no side effects (Figure 1.7).

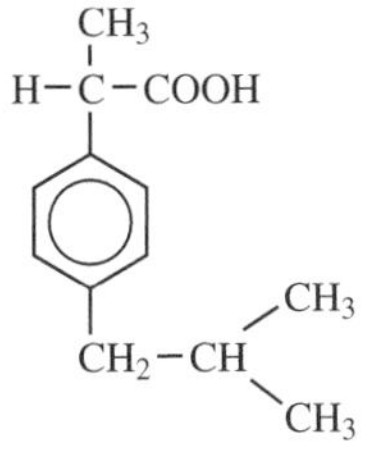

Figure 1.7 Ibuprofen

A similar story to that of aspirin could be told of the discovery and eventual implementation of penicillin and the development of replacements. Mixtures of suitable drugs seem to be a possible answer to combat resistant bacteria – bacteria do not like cocktails!

1.2 Atoms and things

While some Ancient Greek scientists were suggesting medical solutions to common complaints by mixing together natural products, others were 'thinking' and 'wondering' what the composition was of materials in general. Democritus in 400 BCE suggested that all materials were made up of small particles he called atoms. He even invented *symbols* instead of writing the names for elements. In the Western world it was the school teacher and scientist John Dalton, in 1803, who resurrected the idea of the atom. It took until the 1930s for the structure of the atom to be fully understood. *Atoms* are so small that about 1 000 000 000 atoms of iron would fit onto the point of a pin.

Atoms are composed of a heavy central nucleus containing positively charged protons, and these are accompanied by varying numbers of the same-sized neutral particles called neutrons. Rotating in orbits around the nucleus like planets around the sun are negatively charged, very small particles called electrons. The positive charge on the nucleus keeps the negative electrons in place by mutual attraction. The orbits contain only a fixed number of electrons; the inner shell holds a maximum of two and the outer orbits eight electrons or more.

Each element has its own unique number of *protons* and *electrons*. This is called its *atomic number*. Whenever *elements* react together to form *molecules* they try to arrange their outermost electrons to obtain this complete electron shell (of either two or eight electrons), either by sharing electrons with another atom (called *covalent bonding*) or by donation and accepting electrons (called *ionic bonding*). A more complete explanation of these is given in later chapters.

The naturally occurring hydrogen gas molecules, H_2, shares one electron from each hydrogen atom so that each now has a share of two electrons. This is a covalent bond (Figure 1.8). The other method of bonding to get a complete outer electron

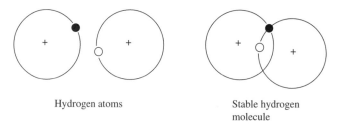

Hydrogen atoms   Stable hydrogen molecule

Figure 1.8   Hydrogen atoms and molecule

shell is demonstrated with sodium chloride or common salt. Here the outermost single electron of sodium is completely transferred to the chlorine atom. Sodium loses an electron so it then has a net positive charge, whereas the chlorine gains the electron and so has a net negative charge. These two oppositely charged particles, called ions, attract each other and form a strong ionic bond (Figure 1.9). A more complete explanation of these is given in Chapter 2 and 7.

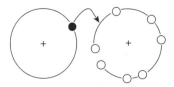

Figure 1.9   Transfer of electrons

As the years progressed, the methods of analysis become more accurate and precise. Scientists were able to detect very small quantities of materials and the structures were worked out. In modern times chemical analysis is done by very accurate and sophisticated techniques. These methods will be discussed in Chapter 11.

1.3 Chemical reactions and the periodic table

Whenever elements and compounds react together to form a stable compound, the atoms always try to rearrange the outer electrons to achieve a complete outer electron shell of two or eight. These complete shells were found to be the structures of the elements in group 8 of the *periodic table*.

The scientists of the nineteenth century discovered new materials that they found to be made up of combinations of simple *elements*. They began to compare the masses of these elements and discovered that this property was a fundamental characteristic of the element – its *atomic mass*.

In 1896 a Russian scientist called Mendeleev found that these numerical values could be put into an ordered pattern which he called the periodic table, which was completed later when more elements were discovered. In about 1932 scientists found that the fundamental property that sequenced the elements in their periodic table order was not their mass but the number of protons in their nucleus. This property is called the *atomic number*, and every element has its own unique atomic number.

In the periodic table according to atomic number all the elements are put in order, each element differing by one unit from its neighbour. It is that simple! (See Appendix 2 for the periodic table.)

The millions of compounds formed by combining these elements together are not so easily systematized. The use of chemical abbreviations and chemical formulae was introduced as some of the molecules were so huge that using names alone for all their contents would lead to impossibly large words. (see *formula* and *symbols for elements* in the Glossary.) There are many millions of compounds made up of approximately 100 different elements. The vast majority of compounds that make up biological tissues are carbon compounds. This branch of chemistry is called *organic chemistry*. There are over a million compounds containing carbon and hydrogen that are arranged into logical groups based upon what is in them and how they react. These groups are called 'homologous series'. Some of these molecules are very large, and proteins are such a group, containing 2000 or more groups of carbon, hydrogen, nitrogen and oxygen atoms. Similarly sugars (or carbohydrates) and fats (lipids) are vast molecules. Of course there are the famous molecules DNA (deoxyribosenucleic acid) and RNA (ribosenucleic acid), which are combinations of smaller groups joined together in their thousands. These molecules are in twisted bundles inside cells, and if they were untwined and strung end to end the molecules in our body would stretch to the sun and back 600 times.

When these protein and other molecules inside our cells are working efficiently then we are well, but if they go wrong, something has to be done. Usually our own body mechanisms can correct these faults itself, but sometimes medication and drugs are needed. That is the beginning of our story about the chemistry of cells and drugs.

Understanding of these complex chemicals needs to be built up in small steps by studying the chemistry of their component parts. Drugs and medicines containing hydrocarbon compounds are covered in Chapter 2; compounds containing OH groups are studied in Chapter 3; the precursors of sugars and fats start with a study of carbonyl compounds in Chapter 4; and the starting point for understanding proteins is the study of amino compounds and amino acids in Chapter 5. Some of the processes involved in the chemistry of medicinal compounds require an understanding of what is meant by covalency, acids, oxidation, solubility, the speed of a reaction and the role of metal ions. All these topics are considered in separate chapters. The growth of analytical techniques and radioactivity are covered in Chapters 11 and 12. Recent chemical and biomedical research is summarized in Chapter 14. Chapter 15 was written to put numeracy into a chemical perspective.

Answers to the diagnostic test

1. Plants   (1)

2. Proton, $+1$; neutron, 0; electron, -1 (3)

3. Sharing electrons   (1)

4. Donating and receiving electrons   (1)

5. Willow tree   (1)

6. Democritus   (1)

7. Periodic table   (1)

8. Alexander Fleming   (1)

Further questions

1. What is the difference between an atom and a molecule?

2. What determines the chemistry of an atom, the outer electrons or the nucleus?

3. What is the name given to particles with positive or negative charges?

4. On which side of the periodic table would you find the metals?

5. The huge branch of chemistry devoted to the study of carbon compounds is called what?

6. What is a homologous series?

7. Aspirin has some side effects, what are they? Name a replacement drug that was developed to eliminate these side effects.

8. What is the difference between atomic number and atomic mass?

References

1. A. Dronsfield. A shot of poison to aid surgery. *Education in Chemistry*, May 2003, 75.
2. J. Le Fanu. *The Sunday Telegraph*, Review, 31 August 2003, 4.
3. *Aspirin*. Royal Society of Chemistry, London, 1998.
4. S. Jourdier. A miracle drug. *Chemistry in Britain*, February 1999, 33–35.

2 Covalent Compounds and Organic Molecules

It is assumed that you have fully understood the principles outlined in Chapter 1.

Learning objectives

- To appreciate the significance of covalent bonding for some biological molecules.

- To write, name and understand the structures of some relevant organic compounds.

- To know what is meant by 'isomerism' and appreciate its importance for metabolic processes.

- To appreciate that there are millions of organic molecules.

Diagnostic test

Try this short test. If you score more than 80 % you can use the chapter as a revision of your knowledge. If you score less than 80 % you probably need to work through the text and test yourself again at the end using the same test. If you still score less than 80 % then come back to the chapter after a few days and read it again.

Chemistry: An Introduction for Medical and Health Sciences, A. Jones
© 2005 John Wiley & Sons, Ltd

1. What is the bonding called where the electrons are shared between atoms?
   (1)

   What is the bonding called where the electrons are donated by one atom and received by the other?
   (1)

2. What is the significance of the structures of helium, neon and argon for chemical bonding?
   (1)

3. What type of bonding holds the majority of the atoms in our body together?
   (1)

4. What materials are made when glucose is burned in air?
   (2)

5. What are the chemical formulae for water, carbon dioxide and ammonia?
   (3)

6. What does 'unsaturated' mean in the phrase 'margarine contains polyunsaturated fats'?
   (1)

7. What do the D and L mean in an optically active isomer?
   (1)

8. Which is the asymmetric carbon atom in lactic acid of formula, $CH_3 \cdot CHOH \cdot COOH$?
   (1)

9. Name the compounds $CH_3 \cdot CH_2 \cdot Cl$ and $CH_3 \cdot CH_2 \cdot CH_2 \cdot CH_3$.
   (2)

10. What is meant by isomerism?
    (1)

Total 15 (80% = 12)
Answers at the end of the chapter.

Alan Baxter, a British skier, lost his winter Olympics bronze medal in 2000 because he used an American Vick inhaler and not a British one.

The British Vick contains a mixture of menthol, camphor and methyl salicylate but the American Vick also contains a further compound, L-methamphetamine.

This is used as a decongestant and has no performance-improving properties, whereas its optically active isomer, D-methamphetamine (commonly known as 'speed'), is a prohibited drug and is a performance improver. However, he was convicted because the Olympic rules and accompanying analysis of materials did not discriminate between the two isomers. All they said was that he had ingested methamphetamine and that was illegal. The chemical explanation of this will be discussed in this chapter. Many believe he was wrongly penalized because of someone's chemical ignorance of the difference between these D-and L-compounds. The following sections start to give the background information of chemical bonding leading up to the chemistry behind the problem of Alan Baxter.

2.1 How to make stable molecules

For separate atoms to combine together to form a new stable molecule, the atoms must form a complete outer electron shell. The completely full electron shell resembles those of the group 8 elements of the periodic table, namely helium, neon and argon. This can be achieved in one of two ways:

- by sharing electrons with other atoms (covalent bonding), e.g. H:H as hydrogen gas, H_2;

- by giving away excess electrons or taking up electrons, forming ions (ionic bonding), e.g. $H^+ Cl^-$.

This chapter will look at the first of these options, covalent bonding.

If you are in doubt about writing simple chemical formula or how to balance an equation, see *formula* and *balancing chemical equations* in the Glossary. If you are uncertain of any symbol for an element then refer to the lists in the Appendices.

2.2 Covalent compounds

Why are covalent molecules so important? The majority of the chemicals in our body are held together by covalent bonds between atoms of carbon, hydrogen, nitrogen and oxygen. Substances like proteins, fats, carbohydrates and water are the building blocks of cells and are all covalent molecules.

2.2.1 Chemical bonding in covalent molecules

You will remember that the elements are arranged in a systematic way in the periodic table according to their atomic numbers (i.e. number of positively charged protons on the nucleus). The first 18 elements are given in Table 2.1 (see Appendix 2 for the full table). When atoms react together and share electrons to form a covalent

Table 2.1 Periodic table showing atomic masses (superscripts) and atomic numbers (subscripts)

1	2	3	4	5	6	7	8
^1_1H							^4_2He
^7_3Li	^9_4Be	$^{11}_5B$	$^{12}_6C$	$^{14}_7N$	$^{16}_8O$	$^{19}_9F$	$^{20}_{10}Ne$
$^{23}_{11}Na$	$^{24}_{12}Mg$	$^{27}_{13}Al$	$^{28}_{14}Si$	$^{31}_{15}P$	$^{32}_{16}S$	$^{35.5}_{17}Cl$	$^{40}_{18}Ar$

molecule, they try to obtain a stable electronic structure. They achieve this by forming a similar stable outer electron arrangement to that found in the elements on the right-hand side of the table. These elements are known as the 'noble gases' (helium, neon, argon, etc.). They have full outer electron shells and are unreactive stable elements.

Helium has two electrons in its outer shell. This inner electron shell is smaller than the rest and is full when it contains two electrons. Neon, with a larger outer electron shell, is full when there are eight electrons in its outer shell, i.e. Ne 2.8, and Argon has two a full outer electron shells, i.e. Ar 2.8.8, etc.

Covalent bonding usually occurs between the elements in the centre of the table, e.g. carbon, and hydrogen or those elements on the right-hand side of the table, e.g. oxygen, nitrogen or chlorine. In the following examples each element achieves the stable outer electron structure of helium, neon or argon.

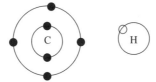

Figure 2.1   Atoms of carbon and hydrogen

The most important set of compounds for us to consider is that of carbon. The most simple carbon–hydrogen compound is methane. Carbon (in group 4) and hydrogen will form methane (CH_4) by a covalent sharing arrangement (Figure 2.1). Carbon has the electronic structure C 2.4 and hydrogen H 1. In its outer shell carbon needs four electrons to achieve the same electronic structure as neon. Hydrogen needs one electron to achieve the same electron arrangement as helium. So if four hydrogen atoms share their electrons with the carbon atom, both atoms will get what they want, namely a full, stable outer electron shell. Methane then becomes a stable molecule (Figure 2.2). (We only need to consider the outside electron shell when electrons form chemical bonds with another atom. It is the outer electrons that 'bang into' each other first and are rearranged when a chemical reaction occurs.) Each electron shell now has the stable arrangement of a group 8 element. Simplistically

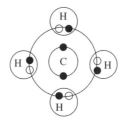

Figure 2.2   Methane molecule

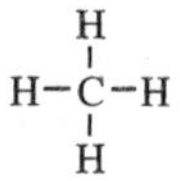

$$H-\underset{\underset{\text{H}}{|}}{\overset{\overset{\text{H}}{|}}{C}}-H$$

Figure 2.3 Methane

we say it is easier for the carbon to share its electrons with another element, like hydrogen, than to try to give away four electrons and become C^{4+} ion, or grab four electrons from another element to become a C^{4-} ion.

Shared pairs of electrons (one electron from each atom) are usually shown as a simple straight line, so methane is drawn as shown in Figure 2.3. There is no such thing as this stick-like formation, but it is a convenient way of representing a pair of electrons between atoms, one from each atom. Now look at another familiar compound – carbon dioxide (Figure 2.4). We write carbon dioxide in its abbreviated form as $O=C=O$ or CO_2. There are two pairs of electrons between each carbon and oxygen atom. We call this a double bond.

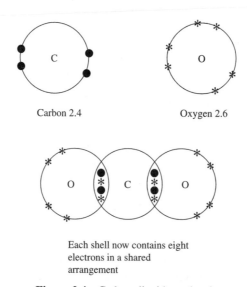

Carbon 2.4   Oxygen 2.6

Each shell now contains eight
electrons in a shared
arrangement

Figure 2.4 Carbon dioxide molecule

2.2.2 Other elements that form covalent bonds

Water, H_2O

Water is also a covalent molecule whose H—O chemical bonds are strong, both in water and biological molecules (Figure 2.5).

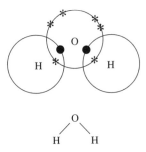

Figure 2.5 Structure of a water molecule

A nitrogen—hydrogen molecule, ammonia

Nitrogen and hydrogen form a stable molecule of ammonia gas (NH_3) in a suitable chemical reaction. Nitrogen has an electronic structure of 2.5 and hydrogen 1. After forming ammonia the hydrogen has a complete shell of two electrons (similar to helium). The nitrogen has achieved a complete eight electron shell by sharing electrons, similar to neon (Figure 2.6).

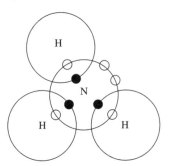

Figure 2.6 Ammonia molecule.

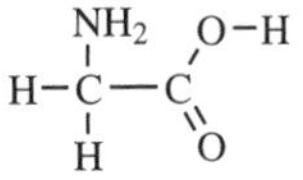

Figure 2.7 Glycine

Amino acids are also covalent molecules, e.g. glycine, $NH_2CH_2 \cdot COOH$ (Figure 2.7). Each line represents a pair of electrons shared between the adjoining atoms. One important property that amino acids have is to allow some of the hydrogens of the O—H group to ionize off, thus making then slightly acid. The cells of our bodies use these small amino acids to form larger protein molecules which also are held together by covalent bonds. These compounds are discussed further in a later chapter.

2.3   General properties of covalent compounds

We and other animals produce their own scents. These are covalent molecules called 'pheromones'; chemicals used to attract or repel people of the opposite sex. Small quantities are produced and can be blown over long distances. Even in very dilute quantities their smell can be picked up. Perfume manufacturers try to copy these smells when making up scents, e.g. 'Musk for men'.

'Hospital smells' are really covalent compounds floating in the air and these attack our noses.

The horrible smells from rotting materials or sewage are also covalent molecules floating in the air. Where there is a smell there are covalent molecules floating around.

2.3.1   Some general physical properties of covalent molecules

All covalent compounds that are liquids evaporate (some solids do as well, and some solid deodorizers depend upon this). This means that the liquid or solid loses molecules from its surface into the air. The more quickly molecules can break out of the liquid or solid, the faster evaporation happens. Covalent compounds (on the whole) have very little attraction between their molecules, but strong bonds within their own molecules. This means that molecules can escape from the liquid fairly easily. Others, like water molecules, evaporate much more slowly because there are stronger bonds holding the molecules together in the liquid (see Section 8.1.1). Gases like oxygen and 'smelly vapours' have little attractive forces between the molecules (Figure 2.8).

Different covalent compounds have their own characteristically shaped molecules with bonds directed at set angles. Some covalent molecules have shapes that 'lock onto' the nerve endings in our nose and produce an electrical signal to our brains. We say that these have a 'smell'. Perfumes, after-shave lotions and scents all contain covalent molecules that produce characteristic smells. So do the less pleasant smells.

Covalent bonds between atoms *inside* a molecule are very strong and a large amount of energy has to be put in to break them, such as burning or strong heat. However, the attraction *between* one covalent molecule and its neighbour is usually quite weak.

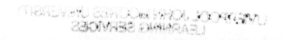

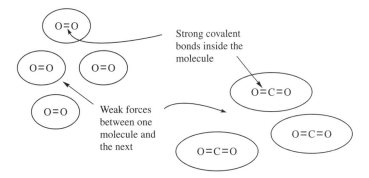

Figure 2.8 Gas molecules

Solubility

Covalent compounds are generally insoluble in water. This is shown by the fact that our proteins, skin and cell materials do not dissolve in the rain! Covalent compounds can dissolve in other covalent liquids like oils or fats. Thus the effectiveness of any medication containing covalent or ionic molecules depends upon their solubility, the type of molecules present in the drug, and the parts of the cells being targeted. Some medications are water-soluble (usually containing ions) while others are fat-soluble (usually containing covalent molecules).

2.4 Characteristic shapes and bond angles within covalent molecules

Inside a covalent molecule, the covalent bonds are directed in space at specific angles. In the case of the methane molecule the carbon–hydrogen bonds are at 109.5°, in other words towards the corners of a tetrahedron with the carbon atom at its centre (Figure 2.9).

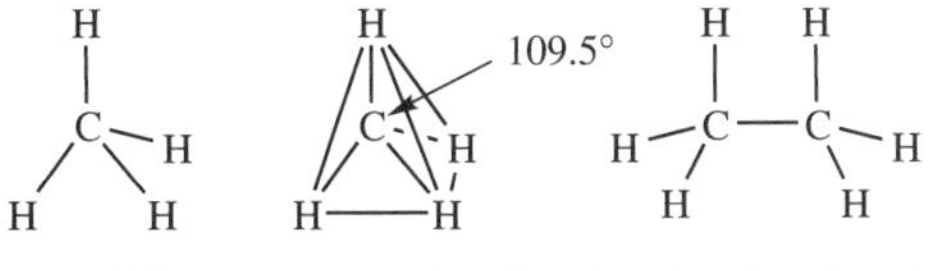

Figure 2.9 Carbon molecules showing bond angles

All covalent bonds have their own characteristic bond angles. These influence the shape of any covalent molecule. The molecules can freely twist about any single bonds, e.g. C—C; H—H; C—H, O—H, but not about any double bonds, e.g. C=O or C=C.

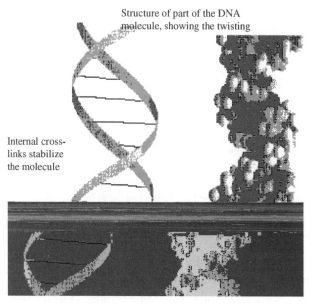

Figure 2.10   DNA chains

More complicated molecules twist and bend to make sure that all the atoms take up positions that allow them maximum free space and non-interference with each other. The very complicated DNA molecules with hundreds of atoms in them twist in a characteristic spiral manner (Figure 2.10). The characteristic shape of each molecule influences its effect on how it behaves in our body and within cells.

2.5   Some covalent bonds with slight ionic character

The majority of compounds containing covalent bonds have very little tendency to split completely to form ions and so they do not conduct electricity. However, there are some covalent molecules that contain groups within them that allow parts of the molecule to form ions. One example of such a compound is the organic molecule of ethanoic acid (acetic acid), which has the formula CH_3COOH. All the bonds within the molecule are covalent. When dissolved in water, however, a few of the O—H bonds in the ethanoic acid break apart and form a small quantity of hydrogen ions and ethanoate ion (Figure 2.11). The majority of the OH bonds do not ionize and none of the other C—C, C—H and C—O bonds of ethanoic acid dissociate to form ions. It is the small amount of H^+ ions that gives vinegar its sharp taste.

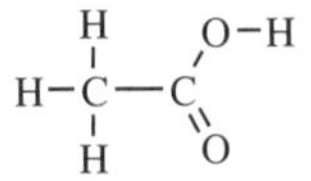

H
|
H-C—C
|
H

O-H
O

Figure 2.11 $CH_3COOH \rightarrow CH_3COO^- + H^+$

2.6 Double-bonded carbon compounds or 'unsaturated' carbon bonds

---

Molecules with double covalent bonds in them, like all other covalent compounds, have characteristic shapes. Carbon dioxide is a linear molecule, meaning that the atoms cannot rotate (or twist) around double bonds. In $O=C=O$ the $O-C-O$ bond angle is therefore 180°. Another example of a compound containing double carbon to carbon bonds is the molecule ethene, C_2H_4 (Figure 2.12). This molecule is flat (or planar), with the $H-C-H$ bond angles at 120° to the $C=C$ bond. The $C=C$ bond cannot twist around the double bond and so the arrangement is quite rigid. The double bond, however, gives the molecule a point of 'vulnerability' for chemical attack. This is because it is under great strain at the point of $C=C$. The 'natural angle' for the $H-C-H$ bonds would be 109.5°, but it is 120° in ethene and 180° in carbon dioxide. These angles distorted from the natural angle of 109.5° put a great strain on the $C=C$ bonds. It is these positions that are the 'weakest links' in the molecule.

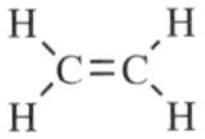

H   H
\\   /
C=C
/   \\
H   H

Figure 2.12 Ethene

2.6.1 Opening up double bonds in ethene

Under the right conditions (usually heat and a catalyst), the double bond of ethane can be opened up. Altering the conditions slightly allows all the ethene molecules to re-join up to form chains of a much larger and more stable molecule called poly- (meaning many) ethene or 'polythene' (Figure 2.13). When the conditions to open up these unsaturated molecules and join them up again into long chains were discovered, the artificial polymer or 'plastic age' started. When a monomer like ethane joins to itself, the process is called 'addition' polymerization.

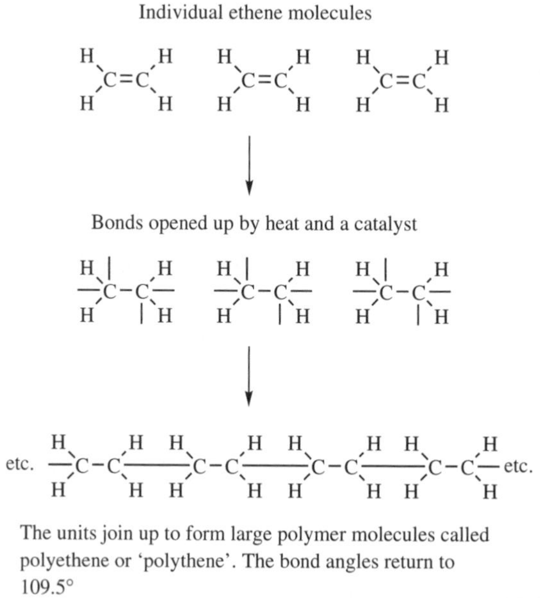

Individual ethene molecules

Bonds opened up by heat and a catalyst

The units join up to form large polymer molecules called polyethene or 'polythene'. The bond angles return to 109.5°

Figure 2.13 Polymerization of ethene

Kathy boasted that she kept her cholesterol down and kept slim with a diet that included a special margarine. 'It never has that effect for me', said Sandy. Mind you, she did have a full cooked breakfast and five pieces of toast and marmalade. The polyunsaturated fat got lost in the crowd of harmful fats.

Generally speaking 'fats' are mostly saturated hydrocarbon compounds. 'Oils', on the other hand, do contain many C=C bonds and are said to be 'unsaturated'. Natural vegetable oils contain many C=C bonds; they are polyunsaturated. 'Poly' means many.

These unsaturated bonds can have the effect of 'mopping up' oxidants made within the body, which can be harmful to cells. Most margarine is made from natural polyunsaturated oils and these are better for healthy living than the saturated fats in butter and cream.

Molecules containing C=C double bonds are said to make a molecule 'unsaturated' whereas C—C single bonds are called saturated bonds. An unsaturated molecule can be 'hydrogenated' to make it saturated by reacting with hydrogen, usually in the presence of a catalyst, e.g.

$$H_2C=CH_2 + H_2 \rightarrow H_3C-CH_3$$

2.7 Some further compounds of carbon

Carbon has by far the greatest number of compounds of any element. The thousands of combinations of carbon with other elements give it the diversity of compounds that makes it the basis of life. Carbon has a reacting power of 4 as it has four electrons in its outermost shell and is placed in group 4 of the periodic table. Its chemical bonds are all covalent. There are over half a million compounds of carbon and hydrogen alone and some are very useful, including the hydrocarbons in petrol.

Compounds of carbon and hydrogen, together with oxygen and nitrogen, make up almost 100 % of the compounds in the cells of our bodies. Without carbon we would not exist. Where does it all come from? It is recycled to us via foods. Green plants obtain their carbon from the carbon dioxide in the air.

The energy that all cells (including ours) require to live comes from chemical reactions between covalent molecules like sugars and oxygen. When carbon compounds react with oxygen, for example, they give out energy and produce carbon dioxide and water. They also make, as a side reaction, a very small quantity of carbon monoxide.

The oxidation of carbohydrates, such as glucose in our body cells, produces carbon dioxide and water. The majority of glucose molecules, for instance, react in the following way when they give out energy:

$$C_6H_{12}O_6 + 6\ O_2 \rightarrow 6\ CO_2 + 6H_2O + \text{energy given out}$$

Carbon dioxide is produced in the cells as the waste product when sugars (e.g. $C_6H_{12}O_6$) are oxidized to give energy. The carbon dioxide is then transferred to the blood, which takes it to the lungs to be exchanged for oxygen. The oxygen is inhaled on breathing and carbon dioxide is exhaled into the surrounding air.

Carbon monoxide is made in very small quantities in our body cells when sugars are oxidized to give energy. It was thought to be of no use in the body and removed as soon as possible. However, in 1992 a startling discovery was made. This carbon monoxide, the potentially poisonous gas, in very low concentrations had an important role. Medical researchers have shown that the regulating role of very small quantities of CO seems to be particularly vital in parts of the brain that control long-term memory. Its complete function in other parts of the body is still being researched. This is a fascinating side of chemistry – the more you find out, the more mysteries are revealed!

'Jodi and Kim killed by a gas from a faulty gas fire in student accommodation', the newspaper headline read. 'Killed by carbon monoxide poisoning' read another newspaper. The high toxicity of carbon monoxide is due to its property of binding to the haemoglobin of the blood. This prevents oxygen from being effectively carried to the cells needing it. The large quantity of carbon monoxide from external sources in this example overloads the blood system and its ability to destroy it. It becomes a killer.

2.8   The carbon cycle

When we breathe out carbon dioxide, we return it to the air for further recycling. Also, when we die our carbon compounds becomes available for recycling. This means that we probably have in our bodies carbon atoms belonging to the dinosaurs, Einstein, our grandparents or even Elvis Presley – what a combination!

Breathing out carbon dioxide means we are contributors to global warming. So don't breathe out if you want to save the world! However, there are worse global warming sources than ourselves, including volcanoes, emissions from cars and lorries, and even 'farting cows' with their methane emission. Fortunately the green plants and trees recycle the carbon dioxide to give us back the oxygen we need, providing we do not cut down too many trees (Figure 2.14).

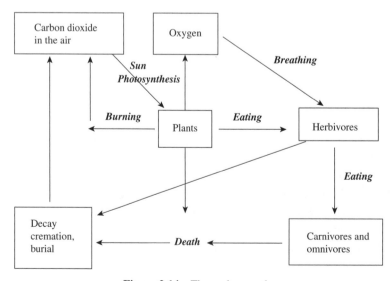

Figure 2.14   The carbon cycle

2.9   Isomerism: some different arrangements of atoms within a molecule

Remember the athlete who lost his bronze medal? Keep reading.

Carbon is one of the most versatile elements. It forms millions of different molecules. Most of these occur naturally in plants and animals. The carbon bonds point at specific angles. This leads to some very interesting molecular variations which are essential to us as humans. Consider a hypothetical carbon molecule that has four different groups attached to a central carbon atom. We will call them A, B, D and E all joined to carbon (Figure 2.15). Remember that carbon covalent bonds are directed at 109.5° to each other. This molecule can be arranged in two ways. These can best be seen if the molecule is drawn in its actual three-dimensional shape. You will then see that, although all the groups are the same, the two forms cannot be superimposed on top of each other; they are mirror images of each other, just as your hands are mirror images of each other. You can put them together, palm to palm, but you cannot put them one on top of another so that they match. We say the two forms are asymmetric. Note that there *must* be four different groups attached to the carbon atom. If any two are the same, then there is only one structure and the forms *can* be superimposed; there is thus no asymmetric carbon atom.

$$
\begin{array}{c}
B \\
| \\
A-C-D \\
| \\
E
\end{array}
$$

Figure 2.15   Asymmetry

It might be worth trying this idea out with, let's say, a tangerine as the carbon, four cocktail sticks as the bonds, and the different groups, A, B, D and E represented by, say, a cherry, a chunk of cheese, a small pickled onion and a piece of pineapple. Now just make sure that the 'bonds' (the cocktail sticks) are as evenly arranged around the tangerine as possible. It is like the carbon at the centre of a triangular prism with its bonds pointing to the corners. If it is not exactly right, don't worry, even an approximate value will demonstrate the principle. Now do the same with another tangerine and another four cocktail sticks.

Sit the two tangerines facing each other on a coffee table (having first cleared off all the magazines and mugs with a deft swipe of the forearm), so that their front two 'feet' are exactly opposite each other. Now stick a cherry on the top cocktail stick of each structure, like A in Figure 2.16. Stick a chunk of cheese on the 'tail' of each, like B in Figure 2.16. Stick a small pickled onion on each of the 'legs' closest to you, like D in Figure 2.16. Stick a chunk of pineapple on each of the 'legs' furthest away from you, like E in Figure 2.16. Get a hand-mirror and place it between the two models, until you are convinced that each is a 'mirror image' of the other. The two forms cannot be put one on top of the other so that all the pieces match up. Try it and see!

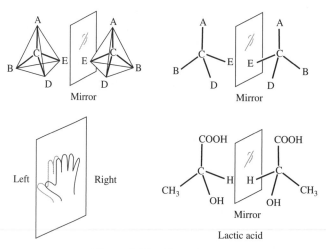

Figure 2.16   Mirror images

Some 'molecules' can be arranged in this way. They are isomers of each other. 'Isomers' mean that the molecules have the same overall formula and contain the same atoms, but in a different arrangement. There are a number of different types of isomerism. The mirror images (Figure 2.16) are examples of a type called 'stereo isomerism' because they are different in their arrangement in three-dimensional space. These molecules are further called 'optical isomers' because it was found that the two isomers twist a beam of polarized light (Figure 2.17) in opposite directions (polarized light is when the light beams are only in one plane and not from all directions). Molecules which have this property have the same 'chemical formula',

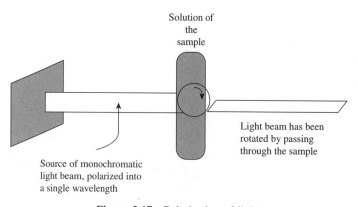

Figure 2.17   Polarization of light

but when a beam of polarized light is passed through a container of each of the two different isomers, one will twist the light beam clockwise (we call this positive or dextro rotatory, D) and the other isomer will twist it anti-clockwise (or negative or laevo rotatory, L). The D and L are often included in the names of such compounds.

All the molecules of sugars, amino acids, proteins will have optical isomers, because there are many asymmetric carbon atoms in their structures that have four different groups joined to a central carbon atom, but the body is very selective of the type of isomer it can use: D-glucose is sweet but its L-isomer is not. The body cells like D-glucose and uses them for cell building and energy sources, but they hate L-glucose.

Consider another actual molecule where this property occurs, i.e. lactic acid, $CH_3 \cdot C^*H(OH) \cdot COOH$. The asterisked carbon atom has four different groups arranged around it (CH_3, H, OH and COOH) and so this molecule must be 'optically active' and have two different isomers. We say the starred carbon atom is 'asymmetric', i.e. not symmetrical. You can draw out the two forms. It is impossible to say which is the D or L form by simply looking at them; they must be tested for their 'twisting properties' with polarized light beams.

2.9.1 General characteristics of 'optically active' isomers

- They must contain an 'asymmetric' carbon atom with four different groups attached to it.

- The two forms cannot fit one on top of the other.

- The two must be mirror images of each other.

- They rotate polarized light in opposite directions.

- They have different physical properties but similar chemical properties.

Look at this amino acid – $NH_2 \cdot CH(CH_3) \cdot COOH$. Can you spot the asymmetric carbon atom? Remember, the three-dimensional 'fruit' model can be used.

After vigorous exercise you can feel a bit 'stiff'. This is because there is still lactic acid remaining in the muscle areas. This causes some pain in the muscles until it is fully worked into the blood and out of the area. If you are feeling stiff, give the area a good rub as soon after the exercise as possible to get the circulation going and remove the lactic acid.

2.9.2 Other isomers

Because there are so many different compounds of carbon, there are many other variations on the theme of isomerism. Most of these are not as relevant to medicine as the ones described here. Some more complicated molecules have many centres of asymmetry in their structures, e.g. some proteins, carbohydrates etc. The body uses many specific molecules. Some cells only accept one type of isomer because the spatial arrangements inside the cell are very tight and only require certain shapes to fit a particular site.

Alan Baxter lost his bronze medal in the winter Olympics in 2000 because he used an American Vick inhaler and not a British one.[1,2] Note the position of the CH_3 attached to the central carbon atom (Figure 2.18). The British Vick inhaler contains a mixture of menthol, camphor and methyl salicylate; the American Vick inhaler also contains L-methamphetamine. This is used as a decongestant and has no performance-improving properties, whereas its optically active isomer D-methamphetamine (commonly known as 'speed') is a prohibited drug and is a performance improver. However, the Olympic Committee does not distinguish between these two isomers in its detective work and chemical analysis. It therefore reported that his urine contained 'methamphetamine' and did not report that it was the ineffective L form. The committee took the medal away from him. If they only had known their chemistry and all about optically active isomers.

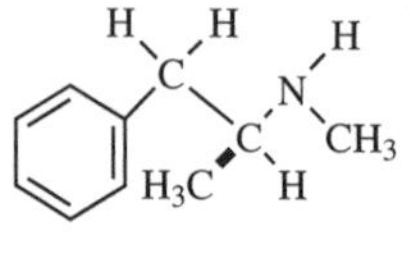

D-Methamphetamine, also known as   L-Methamphetamine, also known as
(S)-(+)-methamphetamine   (R)-(-)-methamphetamine

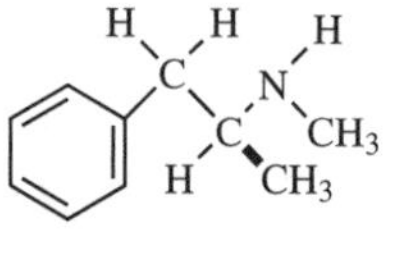

Figure 2.18 The difference between D- and L-methamphetamine

The alternative names for these two isomers in terms of R and S nomenclature refers to their relative positions within the molecule, but the explanation of this system is beyond the scope of this text and it does not add further relevance or explanations to the principles outlined. (See *optical activity* in the Glossary.)

It is important to check the names of compounds on medicine bottles as they might have disastrous effects if wrongly chosen. One common chemical, cholesterol, has 256 possible stereo isomers but only one is actually made in nature.

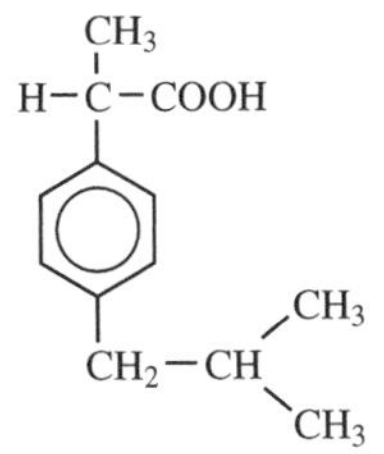

Figure 2.19 Ibuprofen

Ibuprofen, the anti-inflammatory (see Figure 2.19 for its structure), has two isomers but, cleverly, our bodies metabolize the inactive isomer to the active one so a simple mixture of the two forms is sold over the counter. Can you spot the asymmetric carbon atom?

2.10 Naming organic compounds ... if you really want to know!

Because there are so many carbon compounds, there is a great need to be systematic and logical when naming compounds. Chemists use an internationally agreed and logical system. Pharmacists and medics, on the other hand, use more 'historical', ancient and trivial names as these are often shorter and simpler names, but they do not tell you much about the structures. We are going to discuss some of the logical ways of naming organic compounds.

The thousands of 'organic' compounds fall into groups according to the groups of atoms in them. These groups are called 'homologous' series. Each series has a general molecular formula and all the compounds within that series have very similar chemical properties. There are many different homologous series. For simplicity we will discus only the 'alkanes' in detail, as these contain just carbon and hydrogen in their molecules.

2.10.1 General characteristics of any homologous series

- All the members have the same general formula, e.g. alkanes are all C_nH_{2n+2} and alcohols $C_nH_{2n+1}OH$.

- Each member differs from its neighbour by the same increment of CH_2.

- All have similar chemical properties.

- There is a progression of physical properties, e.g. melting points, which increase as the number of carbon atoms increases.

- All the higher members exhibit isomerism.

The thousands of alkanes exist because of the possible arrangement of the carbon atoms either in long chains or in shorter chains but with branching side arms. Some of the values of n can be as high as 200, or even run into thousands.

   When two compounds in any homologous series have the same molecular formula but a different arrangement of atoms or groups of atoms, there is a possibility of a further type of isomerism called 'positional' or 'structural' isomerism. The bigger the molecule, the greater the possibilities of isomerism. Some theoretically possible isomers might not actually exist, due to their impossibly contorted shapes.

2.10.2   Some names and structures

The 'alkanes' have the general formula, C_nH_{2n+2}; all their names end with 'ane' (Figure 2.20). When drawing their structures, the flattened formats of the bonds are mostly used unless it is important to show the carbon bonds pointing at 109.5° into space.

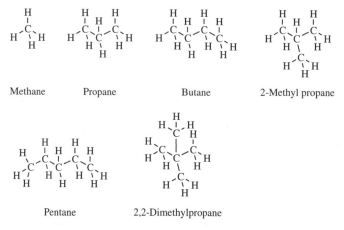

Methane   Propane   Butane   2-Methyl propane

Pentane   2,2-Dimethylpropane

Figure 2.20   Alkanes

   $n = 1$ C_1H_{2+2} or CH_4 is called 'methane' and the CH_3 group is called 'methyl' so CH_3 H is methyl 'ane', which is abbreviated to methane.

   $n = 2$ $C_2H_{2 \times 2 + 2}$ or C_2H_6 is called 'ethane' and the C_2H_5 group is called 'ethyl', so C_2H_5 H is ethyl 'ane' or ethane.

   $n = 3$ C_3H_8 is called 'propane' and the C_3H_7 group is called 'propyl'.

$n = 4$ C_4H_{10} is called 'butane', but there are two possible arrangements of isomers: the ones included in Figure 2.20. These are positional or structural isomers.

$n = 5$ C_5H_{12} is the general formula of a number of compounds called pentanes. Two are shown in Figure 2.20.

Draw out the structure of another isomer of pentane $CH_3 \cdot CH_2 \cdot CH(CH_3) \cdot CH_3$ and give it its systematic name based on the longest carbon chain (see Figure 2.21).

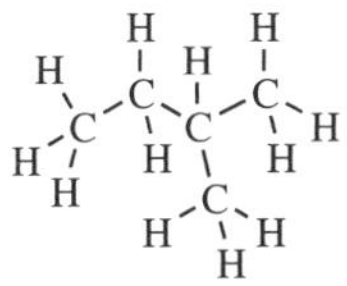

Figure 2.21   2-Methylbutane

2.10.3   Hints on names

You will notice that when naming compounds we try to keep things as simple as possible; the names are always based on the longest continuous carbon chain within the molecule. When groups are attached to the chains their positions are always numbered from the end which gives the lowest numbers. For example, $CH_3 \cdot CH_2 \cdot CH(CH_3) \cdot CH_3$ will be called 2-methylbutane (numbering from the right-hand end) rather than 3-methylbutane (numbering from the left). The molecule does not know the difference – it is up to us to decide. The rule is keep the name and numbering as simple as possible.

2.10.4   Numbering and names

Note that generally we write a hyphen between a number and a letter and a comma between a number and another number in the name (by a worldwide agreement amongst chemists, the names of the formulae are in British format), e.g. 2,2-dimethylpropane.

Recent research

Ethane, a simple hydrocarbon, C_2H_6, seems to be formed by the unusually high concentration of free radicals in patients with a tendency to cancerous growths. These radicals attack proteins and make hydrocarbons, including ethane. A gas detector is being tested to see if the presence of ethane in the breath can be used to detect lung cancer in patients who have shown symptoms.[3]

2.11 Ring structures

A different arrangement of carbon and hydrogen compounds is the rings of three, four, five or six carbon atoms. The most common ring structures are those containing six carbon atoms and are based on a hexagon shape. One such compound is the ring of six CH_2 groups called cyclohexane. Another common compound containing a six-carbon ring is benzene. In this compound there are alternate single and double bonds and this arrangement is called an 'aromatic' ring. When abbreviating these rings it is sometimes convenient to omit the carbon atoms and just show the hexagon shape and the single and double bonds. They were initially called 'aromatic' because they had pleasant smells, but some have recently been found to have pretty vile smells. The molecular arrangement of rings with alternate single and double bonds still retains the terminology of being an 'aromatic structure'. Benzene is C_6H_6 and the group C_6H_5 is called a 'phenyl' group. The ring in C_6H_6, benzene, is flat and planar with H—C—C bond angles of 120°. The C=C bonds do not allow twisting, so the ring is quite rigid (Figure 2.22).

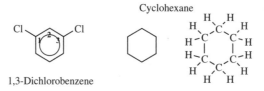

Figure 2.22 Ring structures

These compounds occur widely in nature. When naming ring structures, numbering goes around the ring in such a direction as to keep the numbering as low as possible, e.g. 1,3-dichlorobenzene is preferred to 1,5-dichlorobenzene (numbering in the opposite direction).

2.12 Compounds of carbon containing other groups

Various groups can be attached to any positions of a carbon chain or around the ring. The prefixes which can be added give the names of the groups attached and are as follows:

- Cl, chloro compounds, e.g. $CH_3 \cdot CH_2 \cdot Cl$, chloroethane;

- Br, bromo compound, e.g. $CH_3 \cdot CH_2 \cdot CH_2 \cdot CH_2 \cdot Br$, 1-bromobutane;

- F, fluoro compounds, e.g. $F \cdot CH_2 \cdot CH_2 \cdot CH_2 F$, 1,3-difluoropropane;

- NH_2, amino compounds or amines, e.g. $CH_3 \cdot CH_2 \cdot NH_2$, amino ethane or ethylamine;

- OH, hydroxy compounds or alcohols with 'ol' in their names, e.g. $CH_3 \cdot CH_2 \cdot CH_2 OH$, propan-1-ol or 1-hydroxypropane;

- COOH, carboxylic acids with the C of COOH in the chain are called an '-oic acid', e.g. $CH_3 COOH$, ethanoic acid;

- CHO, aldehydes have 'al' in their names, e.g. $CH_3 \cdot CH_2 \cdot CHO$, propananal;

- C=O, ketones have 'one' in their names, e.g. $CH_3 \cdot CO \cdot CH_3$, propanone;

- C=C, 'unsaturated' groups are shown by names containing an 'ene', e.g. $CH_3 \cdot CH = CH_2$, propene;

- C_6H_5, phenyl, e.g. $C_6H_5 \cdot CH_2 \cdot CH_3$, phenylethane.

If there are two groups of the same type in a molecule, its name has a prefix 'di', e.g. di-chloro-, and you need to say on which carbon atoms the Cl is located, e.g. $ClCH_2 \cdot CH_2 \cdot CH_2 Cl$ would be called '1,3-dichloropropane'. Note the commas between numbers and the hyphen between a number and a letter. More names will be given in later chapters.

2.13 Some further examples with explanations

(1) $CH_3 \cdot CH_2 \cdot CH_2 Cl$ is called 1-chloropropane (the molecule would have been propane if all the atoms were Hs, but one has been removed and replaced with a

Cl on the first carbon atom, counting from either end, hence '1-chloropropane'). Can this compound have any isomers? Try these: $CH_3 \cdot CH(Br) \cdot CH_2Cl$; $BrCH_2 \cdot CH(Br) \cdot CH_2 \cdot CH_2 \cdot CH_2Br$. These Cl compounds are destroyers of our ozone layer if they get into the atmosphere.

(2) In $C_6H_6Cl_6$ the carbons are in a six- membered ring. I will name it, you draw it: hexachlorocyclohexane, or to help you see what is present, hexa chloro cyclo hexane.

Contents of toothpaste

It is improbable that you would be confronted with more complicated names to interpret without specialist advice, but the names on bottles can often contain these long names showing the exact formula of the materials present. These days all food materials should contain a list of contents, but remember what I said about pharmacists and medics: they use their own trivial and nonsystematic names. Large firms, I am sure, do this to confuse the public so that they do not realize the cheapness of a contents inside the fancy package. Whoever would want to call water 'aqua'?

The ingredients list on the side of a packet of toothpaste said it contained:

- aqua – water

- hydrated silica – powdered silicon dioxide

- propan-1,2,3-triol – glycerine etc.

Look at some other packets of materials.

Everyone has their own jargon – don't you? After all you do not want the patients to know what you are talking about. 'He has a cardio-vascular infarction' – why not say a bad heart?

Answers to the diagnostic test

1. Covalent   (1)
   Ionic   (1)

2. Completely full electron shell and stable arrangement   (1)

3. Covalent   (1)

4. Carbon dioxide and water   (2)

5. H_2O, CO_2, NH_3 (3)

6. Contains C=C bonds   (1)

7. D = dextro and L = laevo rotatory   (1)

8. C of CH(OH)   (1)

9. $CH_3 \cdot CH_2 \cdot Cl$ = chloroethane or ethyl chloride and $CH_3 \cdot CH_2 \cdot CH_2 \cdot CH_3$ = butane   (2)

10. Two compounds with the same chemical formulae by different arrangements of its atoms   (1)

Answers to questions in the text

Look at this amino acid: $NH_2 \cdot CH(CH_3) \cdot COOH$. Can you spot the asymmetric carbon atom? Remember the three-dimensional 'fruit' model can be used (Section 2.9.1).

See Figure 2.23.

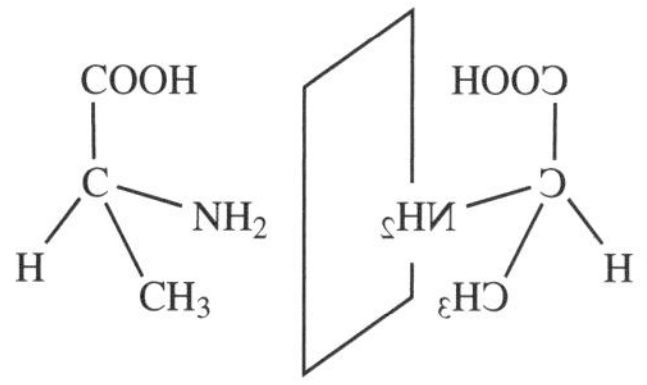

Figure 2.23 Asymmetric carbon atom

Try these: $CH_3 \cdot CH(Br) \cdot CH_2Cl = $ 1-chloro-2-bromopropane and $BrCH_2 \cdot CH(Br) \cdot$ $CH_2 \cdot CH_2 \cdot CH_2Br = $ 1,2,5-tribromopentane (Section 2.13).

I will name it, you draw it: hexachlorocyclohexane or to help you see what is present hexa chloro cyclo hexane (Section 2.13).
   See Figure 2.24.

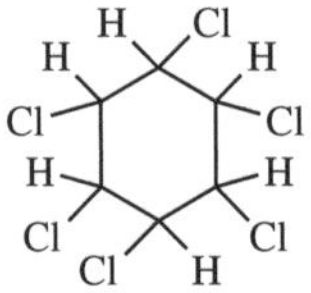

Figure 2.24   Hexachlorocyclohexane

Further questions

If you want to try your hand with other molecules, try these:

1. CH_3COOH, we call it vinegar but what is its systematic name?

2. $CH_3 \cdot CO \cdot CH_3$ the trivial and common name is 'acetone' but what is its systematic name?

3. $NH_2 \cdot CH(CH_3) \cdot COOH$. This is the simple amino acid called alanine. What is its systematic name and is it optically active?

   Name the following:

4. $CH_3 \cdot CH_2 \cdot CH_2 \cdot CH_3$

5. $CH_3 \cdot CH_2 \cdot CH_2 \cdot CH_2OH$

6. $CH_3 \cdot CH_2 \cdot CH(Cl) \cdot CH_2OH$

7. $CH_3 \cdot CH_2 \cdot CH_2 \cdot CH_2(Br) \cdot CH_2 \cdot CH(Cl) \cdot CH_2OH$

8. $CH_3 \cdot CH_2 \cdot C_6H_4 \cdot Cl$. Name this compound. Does it have any isomers?

9. How many isomers are there for the ring structure $C_6H_4 \cdot Cl_2$. What are their names?

10. Which of the structures in questions 4–6 is optically active? Draw out their structures in space to show this.

11. Look at the structure of aspirin in Figure 2.25. Does it have any isomers? Do you think they would necessarily be as active as a pain reliever? Would you say aspirin contains an aromatic ring?

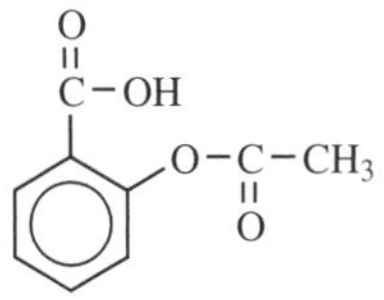

Figure 2.25   Aspirin

12. Draw the structure of 1-methyl-2-chloro-5-bromo heptane.

13. List four characteristics that must be present to make a molecule have a pair of optically active isomers.

14. (i) Draw the spatial structure of 2-chlorobutane, showing the electrons that are present in the covalent bonds. Is this molecule optically active?

(ii) 1-Chlorobutane also has the same chemical formula, C_4H_9Cl, as 2-chlorobutane. These two molecules are also isomers of each other. This is an example of positional or structural isomerism, where the atom of chlorine differs only in the position it occupies on the carbon chain. Draw these two structures and say if there is a further positional isomer of C_4H_9Cl. State your reasons.

(iii) Some chloro hydrocarbon molecules were at one time used as inhaled anaesthetics. A more effective one was 'halothane', which we call 1-bromo-1-chloro-2-trifluoroethane. Draw its structure and write down its chemical formula.

(iv) An early anaesthetic was chloroform (trichloromethane). Write out its structure and chemical formula.

References

1. S. Cotton, Soundbites, more speed, fewer medals. *Education in Chemistry* **39**(4), 2002, 89.
2. S. Cotton, Chirality, smell and drug action. *Chemistry in Britain* **41**(5), 2004, 123–125.
3. On the scent of cancer. *Education in Chemistry* **40**(5), Info Chem. P1, Issue 83, 2003.

3 Organic Compounds Containing Carbon, Hydrogen and Oxygen: Alcohols and Ethers

Learning objectives

- To give an overview of some relevant compounds of C, H and O and particularly the alcohols.

- To prepare the ground for understanding the chemistry of later compounds containing alcohol and ether groups.

Diagnostic test

Try this short test. If you score more than 80 % you can use the chapter as a revision of your knowledge. If you score less than 80 % you probably need to work through the text and test yourself again using the same test. If you still score less than 80 % then come back to the chapter after a few days and read again.

1. What is the chemical name of the alcohol used in alcoholic drinks? (1)

2. The alcohols are members of the monohydric alcohol series with the general formula $C_nH_{2n+1}OH$. Give the formula and names for the alcohols when $n = 1$ and 3 (4)

Chemistry: An Introduction for Medical and Health Sciences, A. Jones
© 2005 John Wiley & Sons, Ltd

3. In what structural way does methanol resemble water? (1)

4. C_6H_5OH has an OH group joined to a six-membered ring. What is the compound called? (1)

5. Glycerine is a trihydric alcohol, $CH_2OH \cdot CHOH \cdot CH_2OH$. What is the chemical name? (1)

6. C_2H_6O is the general formula for an alcohol and also dimethyl ether. Give the formula for each isomer. (2)

Total 10 (80% = 8)
Answers at the end of the chapter.

Ethanol, C_2H_5OH, is commonly called alcohol. Some desperate people drink meths, containing methanol, CH_3OH, as a means of getting drunk but it sends them blind and eventually insane, hence the expression 'blind drunk'. Antifreeze, $C_2H_4(OH)_2$, and glycerine, $C_3H_5(OH)_3$, are members of different alcohol series and behave differently.

3.1 Alcohols, $C_nH_{2n+1}OH$

The compounds commonly known as 'the alcohols' are molecules containing only one OH group; they have the general formula $C_nH_{2n+1}OH$ and should more correctly be called the 'monohydric alcohols', i.e. only one OH group per molecule.

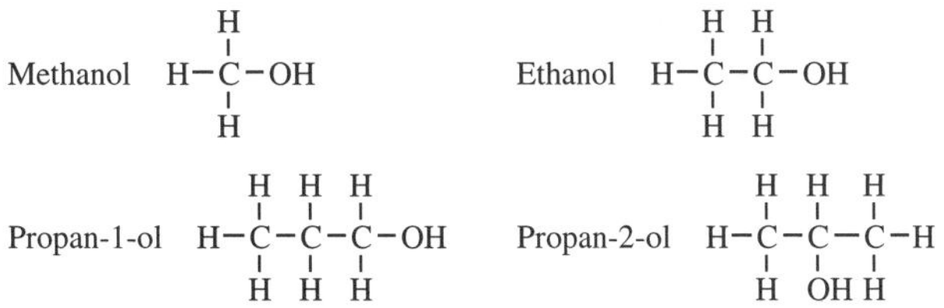

Figure 3.1   Early members of the primary alcohol series

They are famous for the second member of the series, when $n = 2$. This is called 'ethanol', but commonly and incorrectly called just 'alcohol'. All the members have names ending in 'ol' and show the usual gradation of properties shown in any homologous series, e.g. as the molecular weight increases the melting and boiling points increase and the number of possible isomers increases. The alcohol series (Figure 3.1) named as follows: $C_nH_{2n+1}OH$. There is no alcohol with a value of $n = 0$, as this compound would contain no carbon, but interestingly the compound would be $H \cdot OH$, which is water. The alcohol compounds could also be considered to be 'hydroxy alkanes' but they are more usually called alcohols.

$n = 1$ CH_3OH, is methanol; its name is from the name 'methane' CH_4, with an H replaced by 'ol', hence methan ol. It has no alcohol isomers.

$n = 2$ C_2H_5OH, is ethanol. We usually, but incorrectly, call this merely 'alcohol'. It has no other alcohol isomers.

$n = 3$ C_3H_7OH – there are two possible alcohol structures for this formula, propan-1-ol and propan-2-ol.

The occurrence of different isomers in the alcohol series is due to the OH group being located at different positions on the carbon chain. There are no alcohol isomers of the first two members of the series. In C_3H_7OH there are two possible positions of the OH, e.g. $CH_3 \cdot CH_2 \cdot CH_2 \cdot OH$ and $CH_3 \cdot CH(OH) \cdot CH_3$. The alcohols are named using a system that clearly locates the position of the OH (or ol) group on the carbon chain.

$n = 4$ C_4H_9OH is butanol. You can try your hand at drawing and naming the four alcohol isomers. They will all have the same overall chemical formula but different structures (see Figure 3.8, p. 52).

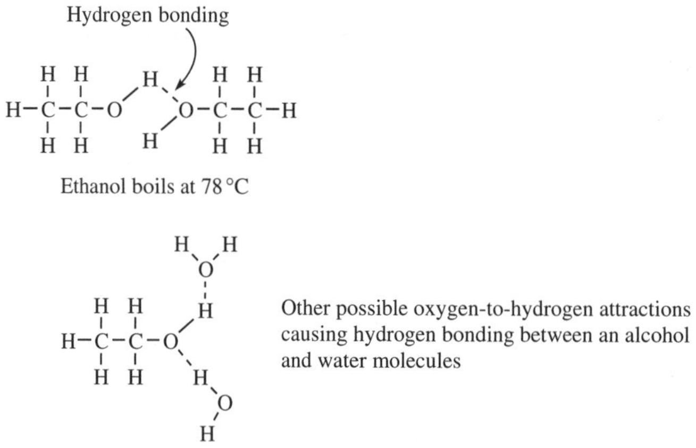

Hydrogen bonding

$$\underset{\substack{H\ H \\ |\ \ | \\ H-C-C-O \\ |\ \ | \\ H\ H}}{}\qquad H \qquad \underset{\substack{H\ H \\ |\ \ | \\ O-C-C-H \\ |\ \ | \\ H\ H}}{}$$

Ethanol boils at 78 °C

Other possible oxygen-to-hydrogen attractions causing hydrogen bonding between an alcohol and water molecules

Figure 3.2   Hydrogen bonding in alcohols

3.1.1 Properties of alcohols

Some of the properties of alcohols resemble those of water, particularly their solvent properties and the chemical reactions of the OH group. They differ most in their toxicity. Only ethanol is non-poisonous in small quantities and low concentrations. Also, the members of the alcohols with lower molecular weights exhibit 'hydrogen bonding'. This is a linking bond between the O of one molecule and the H (of the OH) of another molecule, holding them slightly more strongly together. Hydrogen bonds are not as strong as the normal C—OH bonds or C—C bonds. Hydrogen bonding (Figure 3.2) is discussed in detail in Section 8.1. You can see that there is a linking up of the O and H of two adjacent molecules via a weak bond called hydrogen bonding. This hydrogen bonding makes the boiling points higher than expected because of the extra forces holding the alcohol molecules together and its hydrogen bonding with water if in solution. For example, ethanol boils at 78 °C whereas its molecular weight might lead you to expect it to boil at about 40 °C. Energy is needed to break these hydrogen bonds to convert the liquid into its vapour.

3.2 Properties of alcohols: monohydric alcohols with one OH group

A medical student stole some liquid labelled 'absolute alcohol' to spike some drinks at the Christmas party. Some said ' Phew – that was some potent vodka!' However, its high alcohol content immediately dehydrated the throat and stomach of some of the drinkers, causing much pain. Without treatment some could have died.

Pure ethanol, C_2H_5OH, is called 'absolute ethanol' and is used in medicine as a solvent, specimen storage liquid and cleansing agent. It mixes well with water and can be easily diluted. This property is unusual for covalent organic compounds as they are usually insoluble and immiscible in water. The solubility here is because of the similarity between the OH in the water structure, $H \cdot OH$, and the alcohol C_2H_5OH. Medicinal solutions in alcohol are called 'tinctures'.

Ethanol, we know, is made when fruit sugars ferment. Ethanol (and methanol) can become an addictive material. Prolonged intake can damage tissues, particularly the liver. It has been found[1] that people with alcohol dependency can be treated using a drug called 'antabuse' or disulfiran, which has no effect unless the person consumes alcohol. They then feel sick and nauseous, which is hoped will turn the person off alcohol. The drug acts by inhibiting the normal breakdown, within the body, of one of the oxidation products of alcohol, called ethanal. This builds up in the system, causing the nausea. The mechanism for its action follows the pathway: alcohol, $C_2H_5OH \rightarrow$ by oxidation to ethanal, $CH_3CHO \rightarrow$ further oxidation normally to $CO_2 + H_2O$. In the case of antabuse, this last stage is prevented or delayed.

3.2.1 Fermentation

Everyone knows that fermentation of sugar solutions aided by the enzymes in yeast produces ethanol. This process by the action of fungi breaks down the large sugar molecules in fruits into ethanol:

$$C_{12}H_{22}O_{11} + H_2O \rightarrow 4C_2H_5OH + 4CO_2$$

$$\text{Sucrose} \qquad\qquad \text{Ethanol}$$

The bubbles in champagne are the CO_2 given off in this reaction. Some sugars also give side reactions to produce other compounds. These can be slightly toxic and 'head splitting' in their after-effects. Home brewed wines and beers often contain these unwanted compounds. These unwanted side products are removed by professional wine makers by prolonged storage in wooden barrels, which absorbs the more toxic materials.

Methanol, CH_3OH, is water soluble. 'Meths' is an industrial solvent mixture of ethanol and methanol. It often contains a purple dye and an additive to make it unpalatable to drink. Methanol is poisonous (one spoonful of neat methanol can kill). It can produce similar physiological symptoms to ethanol if consumed in very small quantities in diluted form. It has the additional hazard that it sends people blind and insane and can become addictive.

The alcohols as a group or homologous series of compounds are widely used industrially as a starting point for making other compounds, including drugs, detergents, dyes, plastics etc. The higher alcohols, with three or more carbon atoms, are certainly not consumable. They have unpleasant smells, tastes and are toxic. They are used as industrial solvents for paints and dyes etc. The higher members of the alcohols are waxy materials. Only the first few alcohols are water soluble; the rest are insoluble in water.

3.3 Other alcohols: di- and tri-hydric alcohols

Alcohols that have one OH in a molecule are termed mono-hydric alcohols. Those with two OH groups are called di-hydric alcohols. The compound we use as 'antifreeze' for cooling systems in car engines is such a compound, ethan-1,2-diol (the older name is ethylene glycol; Figure 3.3).

Simple trihydric alcohols contain three OH groups per molecule and an example is propan-1,2,3-triol or glycerol or 'glycerine' (Figure 3.3). Its derivatives and compounds figure greatly in the structures of fats (lipids; see Section 4.8).

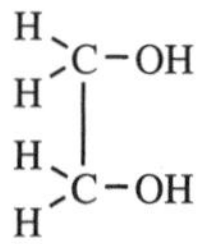

Ethan-1,2-diol
or glycol

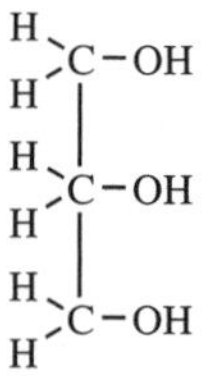

Propan-1,2,3-triol
or 'glycerol'
or 'glycerine'

Figure 3.3 Di-hydric and tri-hydric alcohols

'A common African tree gives hope for AIDS cure', read a newspaper headline. The compound, containing OH groups (Figure 3.4), has potent antifungal

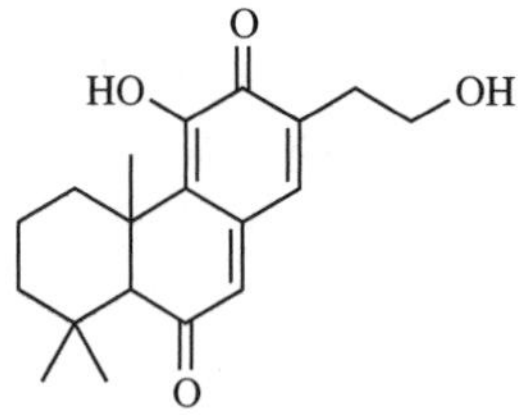

Figure 3.4 A molecule to help in AIDS research

properties stronger than any drug currently on the market. It is being extracted and purified to be used in treating immuno-compromised patients such as those suffering from AIDS.[2] These and other extracts are undergoing clinical trials. Because it takes six trees to make just 50 g of the active compound, a possible synthetic pathway is being researched (note the similarity to the example in Chapter 1 for making aspirin).

3.4 Aromatic OH compounds: phenol

Aromatic compounds have ring structures with alternate single and double bonds. Benzene has a molecular formula of C_6H_6. The carbon atoms are joined together in a hexagonal ring and have alternate single and double bonds.

It is hard to split open the central ring of carbon atoms, but fairly easy to replace (or substitute) the hydrogen atoms hanging onto the carbons of the ring. If one of the hydrogens is substituted by an 'OH' group (hydroxyl group), you get C_6H_5OH. This compound is called 'phenol'. In phenol a single OH group is joined directly onto the ring. A few of the molecules of phenol can release a hydrogen atom when they dissolve in water to form hydrogen ions. This makes the phenol weakly acidic and a strong antiseptic (although not used now on humans). The hydrogens directly attached to the rings cannot be ionized off (Figure 3.5). It is possible to form compounds with more than one OH joined onto the aromatic ring. These are also antiseptics.

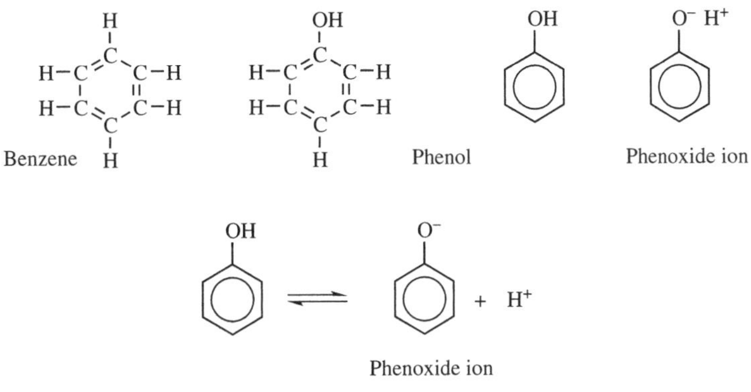

Figure 3.5 Phenol

The phenol derivatives figure in the structures of a number of important reactions, including the synthesis of aspirin and the manufacture of 'phenol – formaldehyde' plastics and glues.

Uncle Jack survived the horrific Battle of the Somme in World War I, but died of antiseptic poisoning by phenol in hospital, where he was treated for an in-growing toe nail by a Florence Nightingale-type nurse. Phenol was discovered in 1834 to be slightly acidic with antiseptic properties (a weak 5 % solution was called 'carbolic acid' and was used to 'kill germs'). When used in the first World War it probably killed many soldiers through overuse, due to its acidic burning effect upon the skin and tissues. More sophisticated antiseptics are now used.

3.4.1 Cyclic alcohols

Cyclic compounds are ring compounds that have *no* double bonds. Cyclic alcohols have at least one OH attached to the ring. The OH here is not acidic and can behave like a usual OH alcohol (Figure 3.6).

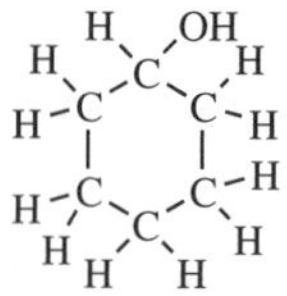

Figure 3.6 Cyclohexanol

3.5 Ethers are isomers of alcohols

Ethanol, C_2H_6O, is normally written showing it as C_2H_5OH, but there is another common material that has the same molecular formula but a different arrangement of the atoms in the molecule (Figure 3.7). This is dimethyl ether or methoxy

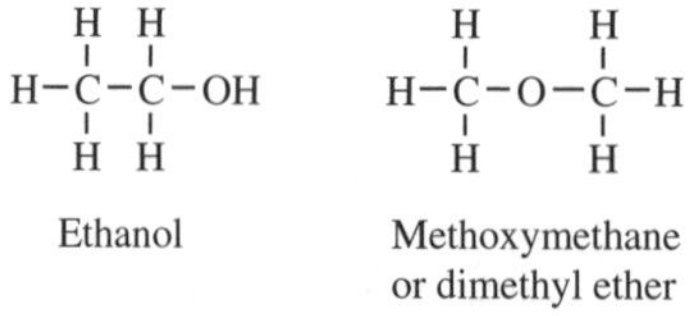

Ethanol   Methoxymethane
                           or dimethyl ether

Figure 3.7 Isomers of C_2H_6O

methane, $CH_3 \cdot O \cdot CH_3$. Alcohols and ethers with the same molecular formula are called 'functional group isomers', since they differ in their functional groups, i.e. OH for the alcohols and -O- for the ethers. The Os in ethers are attached to two carbon atoms on either side.

The ethers are also an homologous series. We will only use the common compound simply called 'ether' as a further example. It is really diethyl ether or ethoxy ethane, $C_2H_5 \cdot O \cdot C_2H_5$. 'Ether' was one of the earliest inhaled anaesthetics and it has a sweet sickly smell. Diethyl ether has a low boiling point (no hydrogen bonding unlike ethanol) and is easily evaporated at room temperatures. If you leave the bottle open it will soon evaporate into a heavy vapour which is extremely flammable. Never use ether near flames or sparks because fire and explosions are possible.

What are the alcohol structures and functional group isomers of diethyl ether, $C_4H_{10}O$ or $C_2H_5 \cdot O \cdot C_2H_5$?

A rock climber become trapped by his arm in a rock fall. To save his life he amputated his own arm with a Swiss army knife without anaesthetic.

It was William Morton, in 1846, who first noted that ether could be used as an anaesthetic for dentistry and surgery. Before that all surgery was conducted on conscious patients. Chloroform, $CHCl_3$, was developed later and Queen Victoria gave the process of anaesthesia credence when she gave birth to a child in 1853 and suffered little or no pain. More recent inhaled anaesthetics include 'isofluorane', $CF_3 \cdot CH(Cl) \cdot O \cdot CHF_2$. Draw its structure (hint: it is also an ether).

Answers to the diagnostic test

1. Ethanol (1)

2. $n=1$, CH_3OH, methanol; $n=3$, propanol, C_3H_7OH (4)

3. Both have an 'OH' in their formula (1)

4. Phenol (1)

5. Propan-1,2,3-triol or 1,2,3-trihydroxy propane (1)

6. CH_3CH_2OH, ethanol and dimethyl ether, CH_3OCH_3 (2)

Answers to questions in the text

n = 4 C_4H_9OH is butanol. You can try your hand at drawing and naming the four alcohol isomers. They will all have the same overall chemical formula but different structure (from p. 46).
See Figure 3.8.

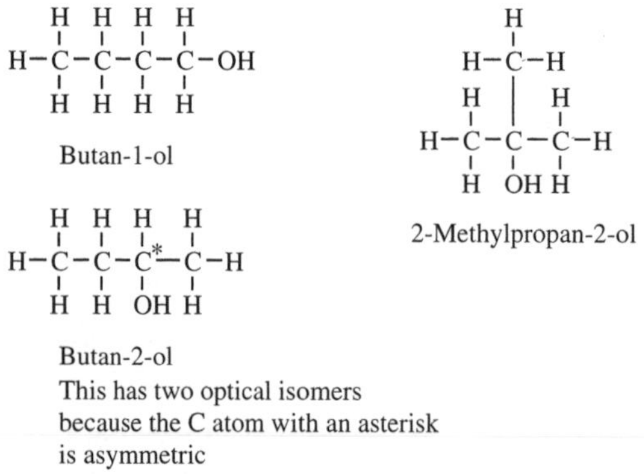

Butan-1-ol

2-Methylpropan-2-ol

Butan-2-ol
This has two optical isomers
because the C atom with an asterisk
is asymmetric

Figure 3.8   The four alcoholic isomers of butanol

What are the alcohol structures and functional group isomers of diethyl ether, $C_4H_{10}O$ or $C_2H_5 \cdot O \cdot C_2H_5$? (from p. 51).

The alcohol structures are the same as given above in the previous question. The ether structures are as show in Figure. 3.9.

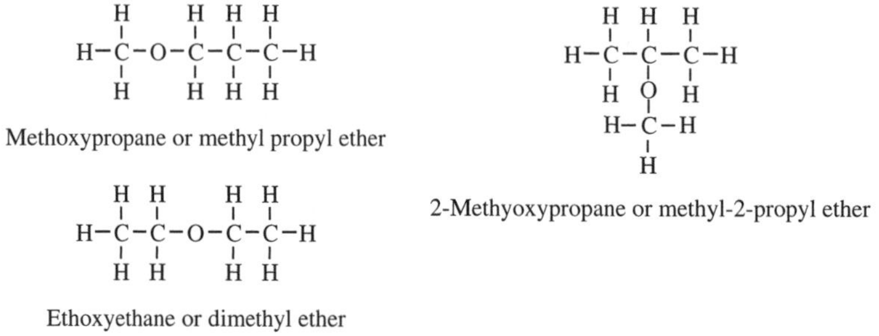

Methoxypropane or methyl propyl ether

2-Methyoxypropane or methyl-2-propyl ether

Ethoxyethane or dimethyl ether

Figure 3.9   Ether structures of diethyl ether

More recent inhaled anaesthetics include 'isofluorane' $CF_3 \cdot CH(Cl) \cdot O \cdot CHF_2$. Draw its structure (from p. 51).

See Figure 3.10

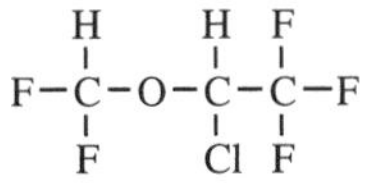

$$
\begin{array}{c}
\quad\ \ H \qquad\ \ H\ \ F \\
\quad\ \ | \qquad\quad | \ \ \ | \\
F-C-O-C-C-F \\
\quad\ \ | \qquad\quad | \ \ \ | \\
\quad\ \ F \qquad\ \ Cl\ \ F
\end{array}
$$

Figure 3.10   Isofluorane

Further questions

1. Explain the two different types of isomerism shown by (i) ethanol and dimethyl ether and (ii) butanol and its four isomers (hint – do not forget the optical isomers of one of them).

2. In what ways are the following compounds similar and in what ways are they different: water, methanol, ethanol.

3. You might expect the boiling points of compounds with roughly the same molecular mass to be roughly the same. They are not. Ethanol C_2H_5OH (molecular mass 46) boils at $78\,^{\circ}C$. Propane C_3H_8 (molecular mass 44) boils at $-42\,^{\circ}C$. Can you suggest why?

4. Which is the first member of the alcohol series $C_nH_{2n+1}OH$ to show:
   (i) optical isomerism;
   (ii) functional group isomerism with an ether;
   (iii) chain or structural isomerism with another alcohol.

5. It is unfortunate that some alcoholics fall to the level of drinking 'meths'. What is 'meths' and how does it affect their health?

6. Give examples and uses of one each of the following:
   (i) monohydric alcohol;
   (ii) dihydric alcohol;
   (iii) a trihydric alcohol;
   (iv) an aromatic compound containing an OH group;
   (v) a cyclic alcohol;
   (vi) an ether.

References

1. S. Cotton. Antabuse. *Education in Chemistry* **41**(1), 2004, 8.
2. Barking up the right tree. *Chemistry in Britain* 2000, 18.

4 Carbonyl Compounds: Compounds Containing C=O Groups

Learning objectives

- To give an introduction to the chemistry of compounds containing the CO (or carbonyl) group.

- To begin to understand the chemistry of carbonyl compounds present in simple molecules and also in sugars, fats and proteins.

Diagnostic test

Try this short test. If you score more than 80% you can use the chapter as a revision of your knowledge. If you score less than 80% you probably need to work through the text and test yourself again at the end using the same test. If you still score less than 80% then come back to the chapter after a few days and read it again.

1. The smell of ethanal is like musty apples. What complaint does this indicate if found on a person's breath? (1)

---

2. Acetone, $CH_3 \cdot CO \cdot CH_3$, is a common ketone. What is its chemical
   name?   (1)

3. Vinegar, $CH_3 \cdot COOH$, is called by what chemical name?   (1)

4. What is a possible ester solvent for glues often abused by glue sniffers? (1)

5. Sugars can contain aldehyde groups; one such common sugar is
   $C_6H_6O_6$. Name this compound.   (1)

6. When two sugar molecules join together, the compound's general
   name is what?   (1)

7. This common compound is taken for angina, headaches and is an
   anti-inflammatory. What could this compound be?   (1)

8. Glucose is oxidized in the body to give energy. What are the waste
   products of this chemical reaction?   (2)

9. Fats or lipids are a combination of two compounds. What are the general
   names for these compounds?   (2)

Total 11 ($80\% = 8$)
Answers at the end of the chapter.

Grace said 'I must have passed out when shopping in the village, because the
next thing I remember is lying here in the doctor's surgery.' The practice nurse
detected a smell on her breath, not of alcohol, but of fusty or rotting apples (of
ethanal). This could be an early indication of diabetes. Later tests showed this
to be true. Suitable immediate treatment was given.

4.1 Simple aldehydes and ketones: carboxylic acids and esters

Carbonyl compounds all contain the C=O group. Compounds that contain carbonyl
groups, C=O, are a part of at least four different simple homologous series, namely

Table 4.1 $\begin{smallmatrix}A\\B\end{smallmatrix}C{=}O$ General formula for all simple carbonyl compounds where A and B can be various groups. A few examples are chosen

Aldehydes, A=H, B=H or any carbon group	Ketones, A and B, both carbon groups	Carboxylic acids, A=OH, B=H or carbon group	Esters, A=OR, R=any carbon group, B=H or carbon group
$\begin{smallmatrix}H\\B\end{smallmatrix}C{=}O$	$\begin{smallmatrix}CH_3\\CH_3\end{smallmatrix}C{=}O$ Propanone or 'acetone'	$\begin{smallmatrix}HO\\H\end{smallmatrix}C{=}O$ Methanoic acid or 'formic acid'	$\begin{smallmatrix}CH_3O\\H\end{smallmatrix}C{=}O$ Methyl methanoate
$\begin{smallmatrix}H\\H\end{smallmatrix}C{=}O$ Methanal	Acetone is an excellent solvent, including for the removal of nail varnish	$\begin{smallmatrix}HO\\CH_3\end{smallmatrix}C{=}O$ Ethanoic acid or 'acetic acid', vinegar	$\begin{smallmatrix}C_2H_5O\\CH_3\end{smallmatrix}C{=}O$ Ethyl ethanoate or 'ethylacetate', the solvent for glues
$\begin{smallmatrix}H\\CH_3\end{smallmatrix}C{=}O$ Ethanal			

aldehydes, ketones, carboxylic acids and esters. These carbonyl groups are also part of important larger molecules of carbohydrates, proteins, sugars and fats or lipids. This covers a large number of molecules used in our bodies for making new cells and producing energy, so understanding the structure and properties of the C=O bond is important (Table 4.1).

You will see from Table 4.1 the following:

- all compounds containing the HC=O groups we call 'aldehydes'; they contain an 'al' in their names;

- compounds containing a C=O group joined to two other carbon groups we call 'ketones'; they have an 'one' in their names;

- compounds containing a C=O joined to an OH are called 'carboxylic acids' and contain 'oic acid' in their names;

- compounds containing a C=O group joined to an O R group are termed 'esters'; These have an 'oate' in their names.

Some common examples of the uses of compounds containing the C=O bonds include the following:

- the smell of ethanal (or acetaldehyde) in the breath of a person is an early indication of diabetes;

- the acid properties of methanoic acid (formic acid) are used as a sting by insects such as red ants and some stinging plants, and so can be treated with a mild alkali, like a paste of sodium bicarbonate;

- the slightly acidic properties of ethanoic acid (acetic acid in vinegar) are used in food preparation and for putting on chips (which are always better eaten out of the paper in a fish and chip shop!);

- ethyl ethanoate or ethyl acetate has a sweet fruity smell and is a solvent for glues; often abused by 'glue sniffers', it acts like a narcotic but probably dissolves part of the sensory cells of the brain, so causing permanent cell damage;

- these carbonyl groups occur in larger and more complicated molecules used in our metabolism and food chains such as sugars, proteins and fats.

It is not the purpose of this book to give the numerous reactions of aldehydes and ketones but only to show their presence in some common biological molecules.

4.2   Carbohydrates, monosaccharides and sugars

Sugars are synthesized by green plants from CO_2 and H_2O in the presence of sunlight. Approximately 200 000 million tonnes of carbon dioxide are taken in by plants from the atmosphere each year. In this process 130 000 million tonnes of oxygen are produced, along with 50 000 million tonnes of organic matter.

The sugars are classified according to following system. Monosaccharides are part of a homologous series with the general formula $C_x(H_2O)_y$. The most common sugars are when x and y are 5 or 6. These are called pentoses ($C_5H_{10}O_5$) and hexoses($C_6H_{12}O_6$). The names of all sugars end with 'ose'.

The aldehyde group is present in some sugars, called 'reducing sugars' because the aldehyde group is a good reducing agent. They all have a CHO group in them. Their general name is aldoses, of which glucose and ribose are the most common members (Figure 4.1). These compounds are aldehyde-oses, abbreviated to aldoses. Reduction and oxidation are explained in Chapter 10. Generally a reducing material will take

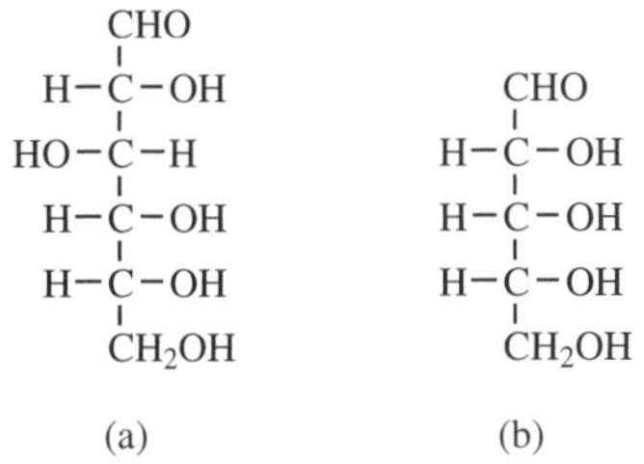

$$
\begin{array}{cc}
\begin{array}{c}
\text{CHO} \\
| \\
\text{H}-\text{C}-\text{OH} \\
| \\
\text{HO}-\text{C}-\text{H} \\
| \\
\text{H}-\text{C}-\text{OH} \\
| \\
\text{H}-\text{C}-\text{OH} \\
| \\
\text{CH}_2\text{OH}
\end{array}
&
\begin{array}{c}
\text{CHO} \\
| \\
\text{H}-\text{C}-\text{OH} \\
| \\
\text{H}-\text{C}-\text{OH} \\
| \\
\text{H}-\text{C}-\text{OH} \\
| \\
\text{CH}_2\text{OH}
\end{array}
\\
(a) & (b)
\end{array}
$$

Figure 4.1   (a) D-Glucose and (b) D-ribose

oxygen away from another molecule and use it to add to its own structure. The CHO group takes on oxygen to form a COOH group or even to break up into CO_2 and H_2O.

The sugar molecules contain many asymmetric carbon atoms. They are therefore optically active and have both D and L isomers. There are 16 optically active aldehyde hexoses alone. Some of them are of value to us in our metabolism and others not. Glucose is the most useful hexose. Our bodies are very selective in what chemicals they use and reject unsuitable ones.

Note that the D and L forms are mirror images of each other (better seen if a three-dimensional model is made of the top three groups and set opposite each other as mirror images). The second carbon atom from the top of the chain is also an important asymmetric carbon atom in glucose. Our bodies are very selective in which isomer they like to use: they use D-glucose and not L-glucose (Figure 4.2).

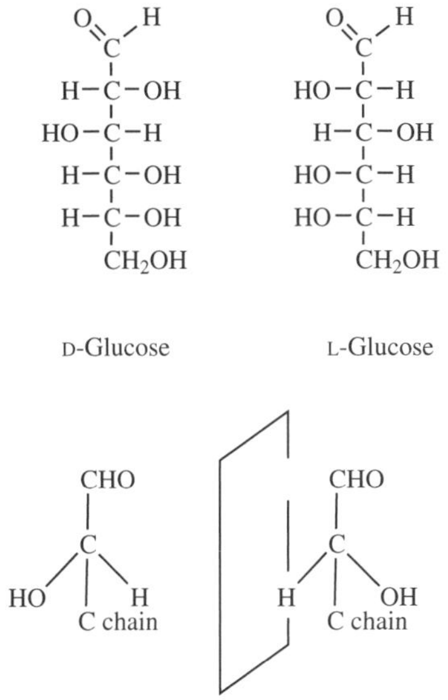

Figure 4.2   Glucose

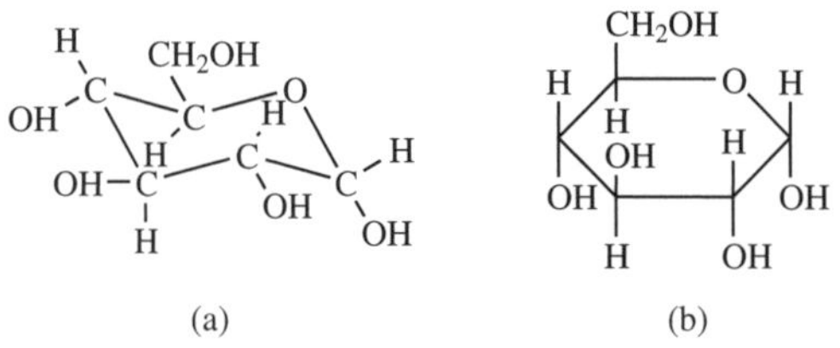

Figure 4.3 (a) Ring structure of glucose showing the true bond angles. (b) Ring structure of glucose as it is often simply written

Glucose, and other sugars, can also rearrange their structure from a linear shape and form a ring (Figure 4.3).

4.2.1 Ketoses

The C=O group can also exist as a ketone, where the C atom of the C=O is joined to two other carbon atoms. These are present in certain sugars with the general name of 'ketoses', of which fructose is an example. Note that aldoses and ketoses can be isomers to each other. Fructose can also exist as a ring structure (Figure 4.4). There is a frightening number of combinations of isomers and structural arrangements for these simple sugars but our bodies are very selective in the ones they require for cell building. You can see that, even when considering these simple molecules, the variations of their structures make them very versatile and important for systems within cells.

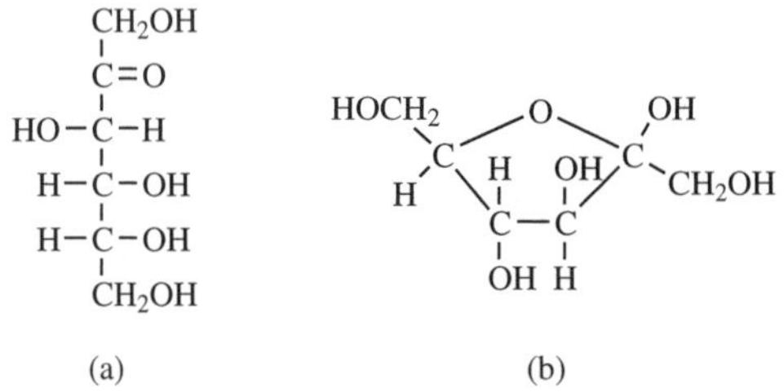

Figure 4.4 (a) Fructose; (b) ring form of fructose

4.3 Disaccharides

Sometimes two sugar molecules join up to form a disaccharide. One well-known disaccharide is called sucrose, $C_{12}H_{22}O_{11}$, formed from one molecule of glucose

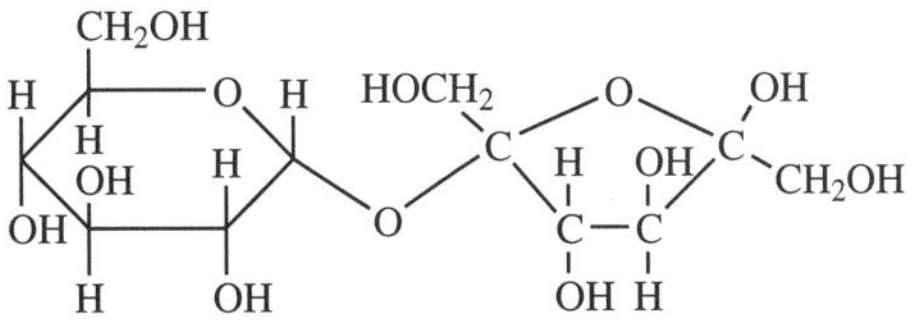

Figure 4.5 Sucrose

joined to one molecule of fructose (Figure 4.5), with the elimination of a water molecule. However, by the action of water in the presence of an enzyme catalyst, these two sugar units can be broken apart. We might put sucrose in our coffee, but the cells of our body cannot absorb large molecules so by enzyme hydrolysis our digestive system breaks down the sucrose to give us the glucose we need.

In more complicated molecules each end of these molecules can join up with further sugar units. When many hundreds of units of carbohydrates are linked together, the compounds are called 'polysaccharides'. Starch and cellulose are such compounds.

4.4   Digestion of sugars

The large molecules of the polysaccharides (starches and cellulose) present in foods are too big for us to digest directly. Our bodies hydrolyse (i.e. action of water in the presence of an enzyme) the starches and break them up into smaller molecules, e.g. glucose. Our systems cannot hydrolyse the complicated long chains of sugar units in cellulose, but those of grazing animals can.

The small glucose molecules are then small enough to be absorbed by the villi on the surface of the small intestine and carried to the liver. The glucose can enter the blood system to be circulated and be readily available to the cells for energy release, for example at muscle endings, as required.

(1) If the body cells require instant energy (as in exercise), the glucose is supplied with oxygen from the blood and is immediately oxidized on the instructions of a facilitating ADP (adenosine di-phosphate) molecule to release energy:

$$C_6H_{12}O_6 + 6O_2 \rightarrow 6CO_2 + 6H_2O + \text{energy}$$

One gram of carbohydrate gives 17 kJ of energy in the process of respiration. The carbon dioxide is removed from muscle endings by the blood and is transported into the lungs for exhalation.

(2) If our body does not need immediate energy the sugars and glucose are converted into a long chain molecule called glycogen in the liver for storage. Some is also stored in the muscles, brain and blood in readiness for instant use. Quick energy release is facilitated by insulin supplied from the pancreas. If the liver glycogen is used up and its quantities run low, our system says it is time to eat again. If we do not eat, the cells start to catabolize (break down) some of the body fats and proteins. Only if we are starving, or eat a series of low-carbohydrate meals, does this begin to have an adverse effect upon the body. Some slimming schemes use this method but they have to be carefully controlled to prevent permanent damage. Long-distance and marathon runners take in high-carbohydrate meals a day or so before the race. This gives them enough energy to prevent breakdown of body proteins.

(3) Glucose can also be used by the cells to synthesize amino acids, which in turn can be linked up to form proteins and are stored.

(4) If all the storage areas of glucose are full, it is used to make fats which can be stored in many places all over the body. Thus we can get 'fat' through eating too many sweet things.

(5) Sometimes when there is excess glucose in a diet it can be excreted in the urine. People with diabetes also have high blood glucose levels and often eliminate the excess glucose into the urine. The urine test is often used as the initial test to see if a person is a diabetic, along with a smell of rotten apples on their breath. This is followed up by other more accurate and specific tests.

4.5   More about sugars – if you really need to know!

Carbohydrates are part of a homologous series, $C_x(H_2O)_y$. This does not clearly show the types of groups that are present. There are carbonyl groups and OH groups present. The 'monosaccharides' start with $x = 3$, but the most common compounds we encounter are those where $x = 5$ or 6. The $x = 5$ contains five carbon atoms; the series is called pentoses. One pentose known as 'ribose' is a vital part of the large molecules of RNA. Deoxyribose is part of the equally important molecule DNA (Figure 4.6).

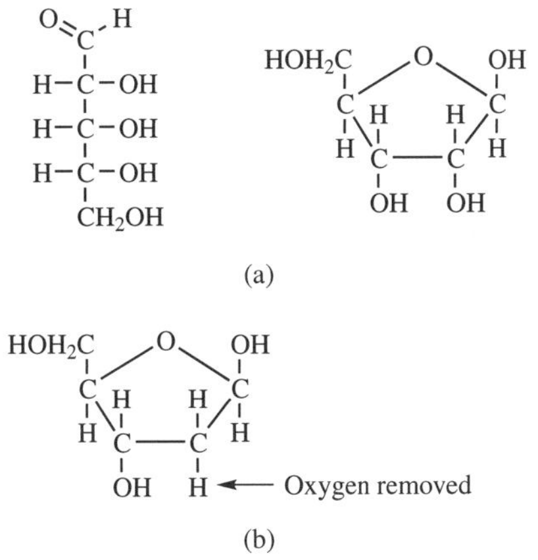

(a)

(b)

Figure 4.6   (a) Two forms of ribose; (b) deoxyribose

4.6   Carboxylic acids: another set of CHO compounds containing C=O groups

These have the general formula of $C_nH_{2n+1} \cdot CO \cdot OH$. All the names of the acids end in 'oic' and the series starts with $n = 0$ (Figure 4.7). The lower acids have the familiar smell of vinegar, but butanoic acid ($n = 3$) smells of rancid sweat.

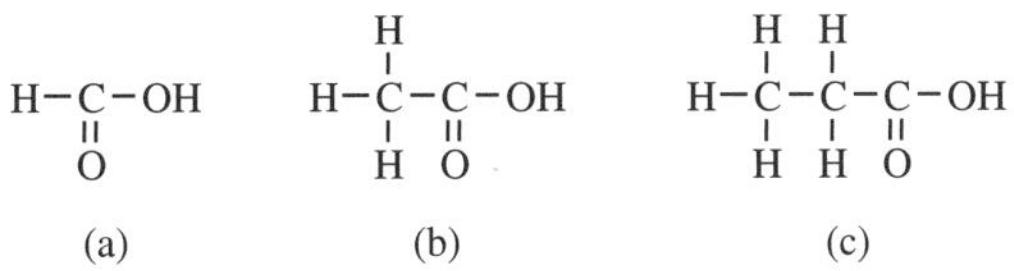

(a)   (b)   (c)

Figure 4.7   (a) Methanoic acid or formic acid; (b) ethanoic acid or acetic acid; (c) propanoic acid

Some members of this series with larger numbers of carbon atoms are called 'fatty acids'. This is because some form part of esters with glycerol, and their compounds are present in fats. Stearic acid, $CH_3(CH_2)_{16}COOH$, is found in animal and vegetable fats. Arachidic acid, $CH_3(CH_2)_{18}COOH$, is found in peanut oil.

4.7   Salts and esters

Like all acids, the carboxylic acids will form salts with alkalis. With sodium hydroxide ethanoic acid will form sodium ethanoate (or sodium acetate). Salts of 'oic' acids form 'oate' salts:

$$CH_3 \cdot CO \cdot OH + NaOH \rightarrow CH_3 \cdot CO \cdot ONa + H_2O$$

With an alcohol the salts are known by the special name of 'esters'.

$$Acid + alcohol \rightarrow esters + water$$

$$CH_3 \cdot CO \cdot OH + C_2H_5OH \rightarrow CH_3 \cdot CO \cdot OC_2H_5 + H_2O$$

$$Ethanoic\ acid + ethanol \rightarrow ethyl\ ethanoate + water$$

Some esters have characteristic fruity smells and are generally used as excellent industrial solvents. Esters like ethyl ethanoate, also called ethyl acetate, are used as a solvents for glues. It is this compound that glue sniffers love. Unfortunately they do not realize that it kills off brain cells and can eventually lead to death.

Selected esters are also used as additives to give fruity smells to certain foods and scents. Some of the exotic scents are often a subtle mixture of various esters.

Got a headache? Take 2 aspirins.

Got a heart complaint, like angina? Take half an aspirin a day.

Got arthritis? Take an anti-inflammatory aspirin.

Got a sore throat? Gargle with a solution of soluble aspirin.

We have probably all taken an aspirin at some time or another. As far back as 2400 years ago ancient records point to people using willow bark as a

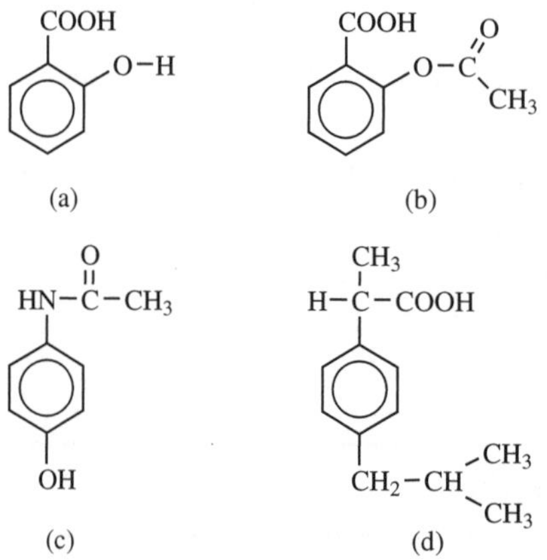

Figure 4.8 (a) Salicylic acid or 2-hydroxybenzoic acid; (b) aspirin; (c) paracetamol; (d) ibuprofen

painkiller. In ancient Britain some complaints and pains were called 'agues'. One common swamp ague was probably malaria. Its effects were eased by chewing willow bark. It has been shown that it contains chemicals similar to modern day aspirin. It was in the 1890s that 'aspirin' was first synthesized artificially. Aspirin is now widely used as a painkiller and also as an effective medicine for reducing the incidences of heart disease. Most of our drugs can be traced back to plant origins (Chapter 1).

Aspirin is an ester of salicylic acid and ethanoic acid. Paracetamol and ibuprofen are also compounds containing C=O bonds (Figure 4.8).

4.8   Lipids or fats

Fats are made from one molecule of the trihydric alcohols, namely glycerol attached to three long-chain carboxylic acids (called 'fatty acids') of various types. This is a tri fatty acid ester of glycerol and is one of a class of compounds called fats or lipids (Figure 4.9).

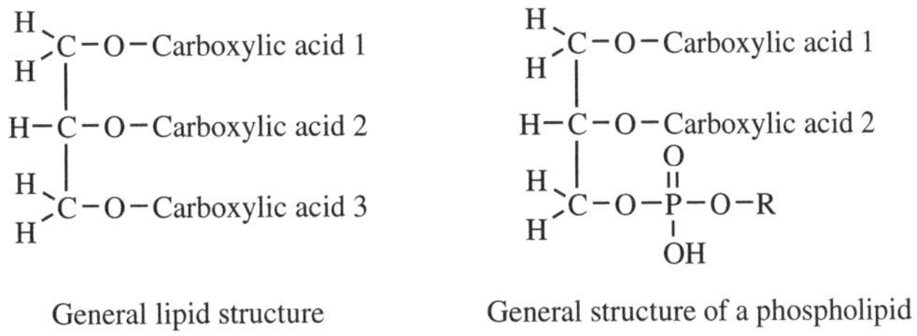

General lipid structure   General structure of a phospholipid

Figure 4.9 Glycerol ester or lipids and phospholipids

Phospholipids (Figure 4.9) are similar to the ester fats but they contain only two fatty acids joined to the glycerol. The other linkage is to a phosphate group. These compounds form a large part of the molecules making up the cell membranes. They have specific properties that allow the passage of different molecules through the membranes. Nutrients are allowed to enter and waste products to leave. This is governed by the mechanisms and structures of these compounds amongst others.

'I'm on a diet to lose weight', said Sandra. 'It costs a fortune.'

'I go to an exercise class to lose weight', said Joseph, 'costs a bit at the gym.'

'I have a light breakfast, a salad snack for lunch and one main meal a day and walk to work. Costs nothing'.

Body fats are the main store of 'energy' in the body: 1 g will give 38 kJ of energy when burned. Some foods, like fatty meats, contain these compounds. Fat is stored in adipose tissues in the cells under the skin. It acts as a protective layer or insulator as well as a supplier of energy. It is synthesized in the body mainly from foods rich in carbohydrates. Fats also dissolve the fat-soluble vitamins A and D.

Saturated fatty acids occur in nature in foods as $CH_3(CH_2)_nCOOH$ where n can have a value from 2 to 20. Fatty acids with high n-values are called 'long-chain fatty acids'. Unsaturated fatty acids are compounds that contain at least one pair of double C=C bonds. One such compound occurring in olive oil and pork fat is oleic acid:

$$CH_3(CH_2)_7CH=CH(CH_2)_7COOH$$

When these unsaturated oils are treated with hydrogen, heated and the vapours passed over a metallic nickel catalyst, double bonds pick up hydrogen and become saturated. The substance becomes more solid. Margarine is made from polyunsaturated oils and is treated to loses some of its degree of unsaturation and make it more solid and 'butter-like'. 'Soft' margarine contains more double bonds than the 'harder' margarine. The unsaturated bonds are thought to be healthier than the more saturated fats occurring in butter.

4.9 Chemical energy in cells

All food contains chemicals. These chemicals are the source of energy. The energy can be released in several ways. The chemical bonds in ATP (adenosine triphosphate) are termed 'high-energy' bonds and the processes of oxidation and reduction are involved in making and breaking these.

Oxidation is a process of electron loss or the uptake of oxygen or removal of hydrogen from a molecule. *Reduction* is the opposite of all these. Whenever something is oxidized then the substance that does the oxidizing is itself reduced (for details of these concepts see Chapter 10). An example of this process are two materials present in cells, lactic acid and pyruvic acid. They undergo an interchange of oxidation and reduction (Figure 4.10). Some of the energy released in this, and many other, oxidation reactions of the body converts low-free energy ADP in the presence of a little more phosphate into high-free energy ATP. This contains high-energy phosphate bonds. The energy stored in the ATP can be transferred to other molecules when required in anabolic processes.

Figure 4.10 Lactic acid/pyruvic acid interchange

The efficient inter-conversion of molecules in the cells to give out, and take in, energy occurs in the presence of other 'carrier' chemicals. These are a basic set of reactions that keep us supplied with energy. These reactions occur smoothly and efficiently within the cell but are difficult to do in a 'test tube' reaction in the laboratory. This shows how well designed the cell processes really are. The wider, more elaborate cycle of events and chemical reactions is called the Krebs cycle (Figure 4.11).

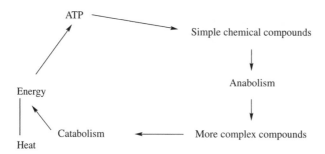

Figure 4.11 The Krebs cycle

4.10   Chemicals in food

The intake of food is essential for the efficient production of energy for all living things. Humans are not like plants, which use the carbon dioxide of the air to harness the sun's energy. We are dependent on the use of plants as food, or other animal material which in turn has already eaten plants. The types of food that are essential for healthy living are mostly carbon-based molecules: carbohydrates, proteins and fats, plus small quantities of minerals and vitamins and water. The effective use of these materials is governed by the interchange of chemical reactions. These are woven into the whole body metabolism. All are geared to give an efficient supply of energy to keep us warm, and for use in growth and cell replacement.

4.10.1   Anabolic and catabolic processes and foods

Chemical reactions within cells that can combine simple substances into more complex or larger molecules are called *anabolic* processes. These reactions require energy (usually heat) to make the reactions proceed and often involve dehydration (removal of water molecules). Such a reaction is the synthesis of large molecules of food protein from small amino acid molecules. The amino acid units have been formed by breaking down food proteins taken in from other foods that have been eaten. These chemical reactions that break down foods are called *catabolic* processes. These reactions involve hydrolysis (reaction with water) and give out energy.

The fine balance of chemical reactions which release energy (food eaten) and those that require energy (the body and cell-building processes) takes place in the metabolic controllers found in the hypothalamus. The energy transfer agent between these processes involves ATP. Most of the energy produced is lost as heat to the environment. Only part of it is available for cell building, hence the need for regular eating.

4.11   Soaps and detergents

My grandmother was quite poor. During World War I soap was scarce and expensive, so she used to make her own soap and sell it to neighbours. She asked them to collect any fat from cooking meat and give it to her. In a large pot on the fire she would boil the fat with 'soda' solution. I can remember the splashing and spitting of the pot. After a few hours this mixture was allowed to cool and the salt solution added and the mixture stirred well. A thick scum formed which was the soap. This was scraped off and squashed into moulds. It was excellent for washing – she also took in washing to eke out the finances. She did not understand the chemistry but she knew the best recipe for success.

Some carboxylic acids are called 'fatty acids' because they come from 'fats' or lipids. The interesting thing is that if the fat is boiled with sodium hydroxide solution it breaks up. The sodium salts of the acids are soft to the skin, lather in water and are called 'soaps'. This breaking up of the fat ester into the soap is called 'saponification'. Soaps contain groups like RCOONa, where R is $CH_3(CH_2)_{16}COO-$ or sodium stearate. Other long-chain acids in soaps are mixtures of palmitic acid, oleic acid, hence the name 'Palmolive'!

Soaps (Figure 4.12) are good cleansing agents because the molecules have a long tail of carbon atoms which dissolve in any greasy material. The COONa end of the molecule is water soluble. The greasy impurities are washed out since the sodium salts are dispersed or emulsified in the water. These large grease/detergent groups are called 'micelles'.

The ability of soap to help 'wet' a surface makes it a good 'surfactant' or 'wetting agent'. The major problem with soap is that, with any calcium ions present, as in

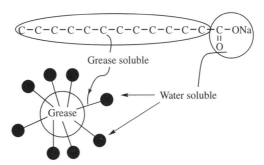

Figure 4.12   Soap

hard water, it forms a scum. This is an insoluble calcium compound of the micelle and dulls the appearance of washed clothes. Because of this, synthetic detergents were synthesized. These are better at keeping the dirt as an emulsion suspended in the washing water and so can be washed out with the water present. Detergents and soaps work in a similar manner. They have a water-soluble group (usually an OH or a SO_3Na) and a 'dirt-soluble' carbon chain.

Answers to the diagnostic test

1. Diabetes	(1)
2. Propanone	(1)
3. Acetic acid or ethanoic acid	(1)
4. Ethyl ethanoate	(1)
5. Glucose	(1)
6. A disaccharide	(1)
7. Aspirin	(1)
8. Carbon dioxide and water	(2)
9. Alcohols and carboxylic acids	(2)

Further questions

1. Carboxylic acids form esters with alcohols. Write an equation for the reaction between ethanoic acid and methanol. The product also has an isomer which is a carboxylic acid. Explain these statements and show the structures.

2. Explain one way the cells of the body get their chemical energy to reproduce and survive.

3. What is meant by an unsaturated fatty acid and why is it more healthy to eat compounds containing these than saturated fatty acids.

4. Give common uses for the following compounds:

   (i) methanal, whose solution in water is sometimes called formalin;
   (ii) ethanoic acid;
   (iii) ethyl ethanoate;
   (iv) propanone;
   (v) glucose;
   (vi) sodium stearate.

5. C_3H_6O is a carbonyl compound that has no reducing properties and is often used as a nail varnish remover or industrial solvent. It has an isomer that is a good reducing agent and is itself oxidized to a carboxylic acid CH_3CH_2COOH. Explain these statements and identify the compounds.

6. Explain how soaps and detergents remove dirt from a material during washing.

7. Explain the difference between anabolic and catabolic processes.

8. Explain how the disaccharide sucrose can be converted into two monosaccharides.

9. Explain the role of ATP and ADP in the release of energy within cells.

10. Look up in *Encarta* or a similar database the discovery of penicillin and paracetamol.

5 Organic Compounds Containing Nitrogen

Learning objectives

- To introduce the chemistry of carbon compounds containing nitrogen, e.g. amines.

- To introduce some structures of amino acids and proteins.

- To show the basic structure of some units of DNA and RNA.

Diagnostic test

Try this short test. If you score more than 80 % you can use the chapter as a revision of your knowledge. If you score less than 80 % you probably need to work through the text and test yourself again at the end using the same test. If you still score less than 80 % again then come back to the chapter after a few days and read again.

1. Ammonia is NH_3. If one H is replaced by a CH_3 group what is the name and formula of the compound? (2)

---

2. The compound in (1) is part of a series called primary amines. What is the general name of the compound if a further hydrogen is replaced by a carboxylic acid group, COOH? (1)

3. Glycine is $NH_2 \cdot CH_2 \cdot COOH$ and is an amino acid. Show how this compound reacts (i) with HCl acid and (ii) with NaOH, an alkali. $(2 + 2)$

4. Show how a dipeptide is formed from two glycine molecules joining together with the elimination of water. (2)

5. A protein is made up of many amino acids joined together. Proteins are broken open by the action of water. What bonds are broken open? (1)

Total 10 (80 % = 8)
Answers at the end of the chapter.

Trinculo, in the *Tempest*, act 2, scene 2, by William Shakespeare, said:

What have we here? Man or fish? dead or alive?
A fish, he smells like a fish … a very ancient and fish-like smell …
A kind of not-of-the-newest poor-John

A nitrogen-containing compound, trimethylamine, is produced in a complaint called 'fish odour syndrome', where the sweat, breath and urine all smell of rotting fish. It is caused by a metabolic liver malfunction and releases trimethylamine in the bowel and gut. Drugs and dietary control can cure the condition.

5.1 Amines and amino acids

Amines are a series of compounds containing carbon, hydrogen and nitrogen. They are part of the building blocks for the important protein molecules involved in metabolic processes. Amines are compounds similar to ammonia, but have one or more of their hydrogens replaced by a carbon chain, e.g. CH_3. When one hydrogen of ammonia is replaced, the series is called the primary amines (Figure 5.1); with two Hs replaced, the series is called the secondary amines, and with all three replaced, tertiary amines.

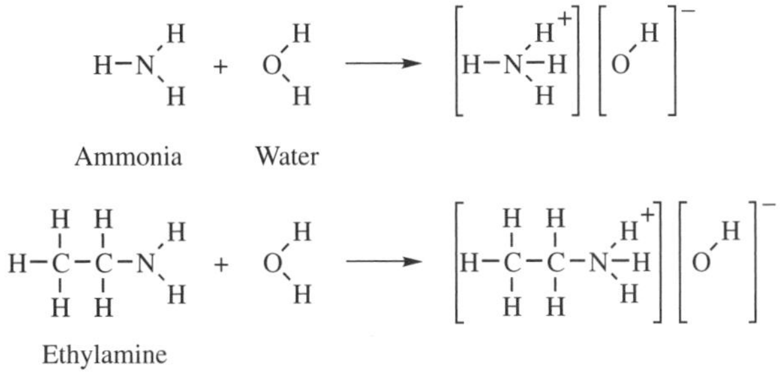

Figure 5.1. Amines reacting with water

The primary amines, $C_nH_{2n+1}NH_2$ are:

- $n = 0$, ammonia, $H \cdot NH_2$ (this is not an amine because it does not contain a carbon atom but you can see the similarity to an amine);

- $n = 1$, methylamine (or amino methane), CH_3NH_2;

- $n = 2$, ethylamine (or alternatively amino ethane), $CH_3 CH_2NH_2$ etc.

Secondary amines are of the type, $(CH_3)_2NH$, dimethylamine, where two Hs are replaced by a carbon group such as CH_3. Similarly $(CH_3)(C_2H_5)NH$ is called methylethylamine and $(C_2H_5)_2NH$ called diethylamine.

Tertiary amines are of the type $(CH_3)_3N$, trimethyl amine, where three Hs are replaced by a carbon group such as CH_3. They smell of fish, as mentioned in the Introduction.

Like ammonia, amines are alkaline in solution with high pH values. The tertiary amines are the most basic or alkaline and primary amines the least basic. They all have a fishy smell resembling ammonia. There is a compound made in the body as a breakdown product of proteins and amino acids and passed out in urine: it is called urea. This is also an amine, with the formula $CO(NH_2)_2$ (Figure 5.2).

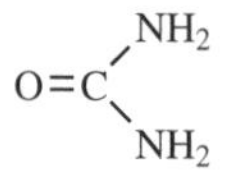

Figure 5.2. Urea

Primary amines are introduced here because they are one of the 'root' compounds of the important series of compounds called the 'amino acids'.

5.2   Amino acids

If one or more of the hydrogen atoms on a carbon atom of an amine is replaced by a 'carboxylic acid group', COOH, as in $NH_2 \cdot CH_2 \cdot COOH$, then the series is called an 'amino acid' because its members contain both an amine group and a carboxylic acid group.

This unusual type of molecule is essential to all living things. It has a basic group (i.e. NH_2) at one end and an acidic group (i.e. COOH) at the other end. This means that the molecules can have dual basic and acidic properties, depending on what environment it is in. This is an important property for molecules so vital for our body cells, which are subject to many changes of acidity.

The acidity of a solution is more quantitatively defined as a pH value. This is a term that expresses the hydrogen ion concentration on a 1–14 scale. Solutions with pH 1–6 show acidic properties, pH 7 is neutral and solutions with pH 8–14 are basic. The lower the acidic scale is, the more acidic the solution; alternatively, above 7, the more basic a solution is, the greater the value.

As the pH of a solution is altered, the following changes occur in amino acids. Starting as a neutral non-charged amino acid molecule, suppose more acid (H^+) is added. It changes from an electrically neutral molecule to a positively charged ion, $NH_3^+CH_2$ COOH. If now it is subjected to an alkaline environment due to OH^- ions being present, then a doubly charged ion (both positive and negative charges are present on the molecule), $NH_3^+CH_2$ COO^-, is formed, and eventually a negatively charged ion, $NH_2CH_2COO^-$ (as H is removed from the NH_s^+ group). The reverse set of formulae would occur if an acid (H^+) was added to the $NH_2 \cdot CH_2 \cdot COO^-$ ion

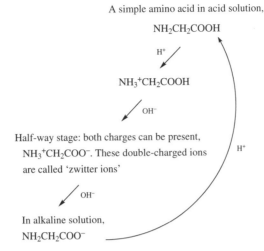

Figure 5.3. Amino acids in acid and alkali solution

(Figure 5.3). These cycles of changes help to maintain the best working pH for the body cells, which are made up of proteins and amino acids.

In equation form:

$$NH_2 \cdot CH_2 \cdot COOH + H^+ \text{(from an acid)} \rightarrow NH_3^+ \cdot CH_2 \cdot COOH$$
$$NH_3^+ \cdot CH_2 \cdot COOH + OH^- \text{(from an alkali)} \rightarrow NH_3^+ \cdot CH_2 \cdot COO^- \qquad (5.1)$$
$$NH_2 \cdot CH_2 \cdot COOH + OH^- \rightarrow NH_2 \cdot CH_2 \cdot COO^-$$

The charged amino acid molecules make excellent controllers of the acidity of an environment, in cells, and in body fluids, including in the stomach, as they can 'soak up' excess acid or alkali. The protein structures on the lining of the stomach help to stabilize the acidity in it and to control the pH. These pH values can often be disturbed by over-eating or -drinking.

Such compounds are called buffers which are discussed in more detail in Chapter 9.

5.3 Peptide formation and protein synthesis

Two or more amino acids can be 'condensed' (i.e. a water molecule removed and the remaining molecules joined together) to form 'peptides' (Figure 5.4). This reaction can be performed in a test tube, but is occurring all the time in the cells of living

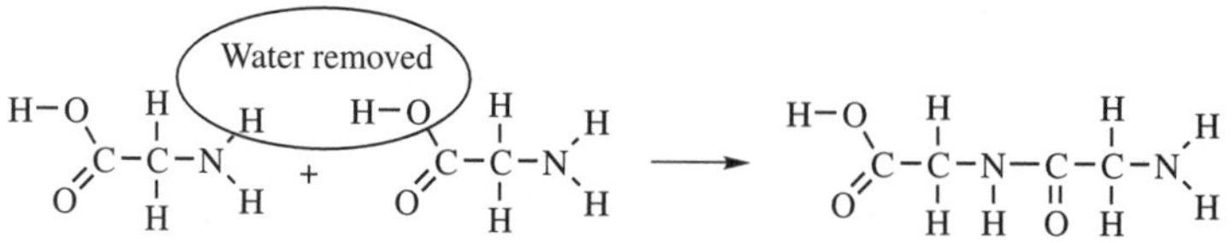

Figure 5.4.   Condensation of amino acids to form a peptide link

things, including us:

$$HOOC \cdot CH_2 \cdot NH_2 + HOOC \cdot CH_2 \cdot NH_2 \rightarrow H_2O + HOOC\,CH_2 \cdot \mathbf{NH\,CO} \cdot CH_2 \cdot NH_2$$

$$(5.2)$$

The CO·NH group is called a 'peptide group' or 'peptide link'. When many of these groups join together to form one big molecule it is called a 'polypeptide' or protein molecule.

Amino acids are invaluable because they can form long-chain polymers needed by our bodies in definite and prescribed patterns as determined by the cells. There are sometimes hundreds of amino acid units in these chains. There are special parts within the cells which synthesize the specialized proteins required from the small units of amino acids.

5.4   Hydrolysis (action of water) of peptides

The opposite of peptide formation can occur in a different environment. Here the break-up of the protein's peptide bond is achieved by the addition of a water molecule, often in the presence of an enzyme catalyst and usually in an acid solution. This type of break-up of the protein is called 'hydrolysis' (Figure 5.5).

$$H_2O + NH_2 \cdot CH_2 \cdot \mathbf{CO} \cdot \mathbf{NH} \cdot CH_2 \cdot COOH \rightarrow NH_2 \cdot CH_2 \cdot COOH + NH_2 \cdot CH_2 \cdot COOH$$

$$(5.3)$$

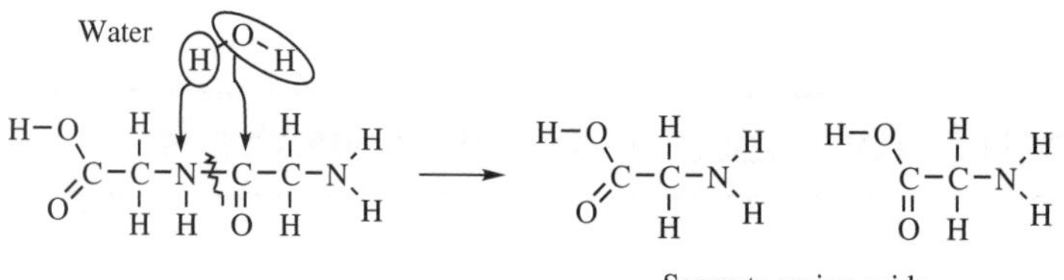

Separate amino acids

Figure 5.5.   Protein chain attacked by the water molecule breaks open to make the individual amino acids

The strong hydrochloric acid, naturally present in the stomach, hydrolyses (adding on the elements of water) the foods containing proteins to form small amino acids for transport by the blood into cells. These small amino acid units are the 'starter building blocks' for specific sequences of amino acids needed inside the cells for protein building. The proteins manufactured in the cells are different from those taken in as food, which is why they first have to be broken down into small units for eventual remaking and rearranging into those specifically required by the cells.

There are approximately 20 separate amino acids that form various proteins of immense chain length, some with as many as 2000 units of different combinations in their chains. Each protein has its own characteristic shape: some are long chains, some are spirals and some are more ball-like. Their shape often determines the specific and characteristic properties of the particular protein.

5.5   Other properties of amino acids

Each protein operates best at a specific pH and temperature If these alter too much it loses its molecular shape and some of the peptide bonds break up. It is then said to be 'denatured'. You will know that an egg white protein is transparent and runny, but when heated in a frying pan it hardens and becomes more rubbery. The egg's proteins have been 'denatured'.

> 'Wonder Power' the unique washing powder with added enzymes that chew up dirt'. The advert says, 'Don't boil the washing when using Wonder Power enzyme'.
>
> 'Enzymes' in the presence of water can attack and break apart proteins and biological body waste products and dirt. Why is it advised not to boil the washing? The enzyme is a protein and can be deactivated or denatured if heated too strongly.

5.6   Protein metabolism

During digestion, the proteins present in food are too large to pass directly through cell walls and membranes so they are broken down into the more simple amino acid units (by enzyme hydrolysis). These small amino acids are absorbed by the villi of the intestines and taken into the blood and the liver. Amino acids are not 'stored' for future use but are used very soon after their production to make cell material, peptides and proteins.

The synthetic process is stimulated by a 'growth hormones' and also by insulin. Many different proteins have different jobs. Some act as enzyme catalysts for other chemical reactions in the fluids and cells of the body. Some go to make haemoglobin, which is synthesized in the bone marrow; others go to make material for muscles, hormones, collagen, elastin, etc.

Proteins are so valuable that they are not normally excreted from a healthy person but are converted into sugars (carbohydrates) by glycogenesis or into fats (lipids) by lipogenesis. Unused material is broken down again into small amino acid units ready to be remade into new protein material. The body has a great way of recycling the 'worn out' but valuable protein material for remaking amino acid units. Amino acid residues that are not needed are usually converted into urea, $CO(NH_2)_2$, before being excreted in the urine.

There are approximately 20 amino acids in use in our body, but 10 of these are the 'essential' amino acids and these must be taken in from food as our bodies are not able to synthesize them. The 10 nonessential, but still important, amino acids can be synthesized from proteins. These amino acids are synthesized by the cells of the body by a process called 'trans-amination', i.e. the body uses the amino groups from unwanted proteins to make the ones it needs.

5.7 Nucleic acids, DNA and RNA

Nucleic acids are large polymers containing three main types of units joined together. These are the nitrogen-containing bases, sugar components and the phosphoric acid part. They occur in all living things and are essential in making proteins and in determining hereditary characteristics. They contain the individual's genetic code, which dictates the characteristic sequence of amino acids in a protein chain.

Deoxyribose nucleic acid (DNA) and ribonucleic acid (RNA) are compounds found in the nucleus of cells of all living things and have the important roles of telling cells what molecules to synthesize or destroy and when. The base groups are selected from the four nitrogen-containing compounds, ardenine (A), guanine (G), thymine (T) and cytosine (C), and these are joined to the 'ribose' (or deoxyribose), a five-carbon carbohydrate or sugar molecule (Figure 5.6). Phosphoric acid groups are also an essential part of the structures of DNA and RNA (Figure 5.7). The units are

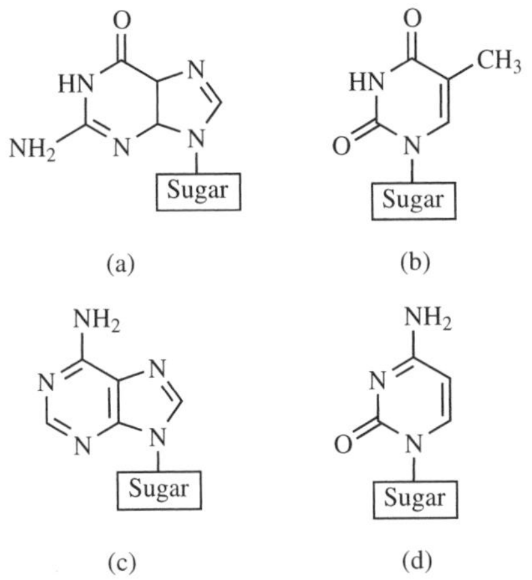

Figure 5.6. (a) Guanine, G; (b) thymine, T; (c) adenine, A; (d) cytosine, C

Figure 5.7. (a) Ribose shown in ring structure form; (b) phosphoric acid

joined up to form very long twisted spiralling chains forming inter-twining helix-like structures. DNA is a double helix (Figure 5.8).

A detailed study of DNA and RNA is outside the scope of this book, but the discovery of the exact structural arrangement was first worked out by Crick, Watson,

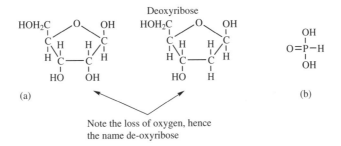

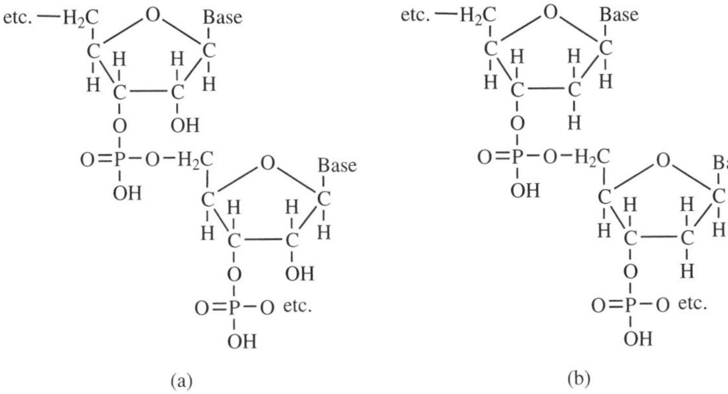

Figure 5.8. (a) A nucleic acid of ribose, base and phosphate; (b) a nucleic acid of deoxyribose, base and phosphate

Wilkins and Rosalind Franklin. The discovery of the structure of DNA makes fascinating reading. The special issue of *New Scientist* of 15 March 2003 to mark the 50 years since the discovery of the structure of DNA contains many interesting details.[1]

Did you know that if you extracted all the DNA from your cells and put them end to end, they would stretch to the sun and back 600 times? This is because we have approximately 10 trillion cells in our body and each cell contains thousands of DNA molecules. These cell molecules are under constant chemical and environmental attack and so there is a similar number of repair events to restore these structures. There are approximately 10^{20} harmful attacks on the cells of our bodies each day from chemicals, oxidizing free radicals, uv light, cigarette smoke, etc. Unless repair is done quickly, these cells can form deformed structures and cause many molecular-based diseases, including cancers. This is why a constant supply of food in a balanced diet is essential for healthy living. Snack food and slimming diets sometimes lack essential proteins and minerals.

'Did you hear there was an attempted rape by a masked man on the campus last night? Fred, along with all the other male students, was asked to supply a sample of hair or skin for DNA analysis. What do you suppose that will tell the police?'

Some parts of the sequence of these long chains are characteristic of the person who made them and cannot occur in anyone else in this exact sequence. Hence, DNA sampling is used to characterize a person. All that is needed is a small piece of hair, a flake of skin or a minute drop of body fluid for analysis to show up this sequence (see also Chapter 11).

The two twisted strands of DNA are held in place by 'hydrogen bonding', shown in Figure 5.9 as lines between the C–G, T–A, etc. groups protruding from the DNA chain. Refer to Chapter 8 to see the mechanism by which hydrogen bonds operate. Here are two examples showing the hydrogen bonding between O and H atoms and also N and H atoms. Without these hydrogen bonds the helix would untwist and we would fall apart – so make sure you get your DNA in a twist when reading this!

5.7.1 A recent discovery

Two scientists working in Baltimore discovered what they called the 'master protein, HIF', which controlled the growth of new blood and oxygen levels in the red blood cells, so essential for curing tumours, cancers and helping recovery after heart attacks.[2]

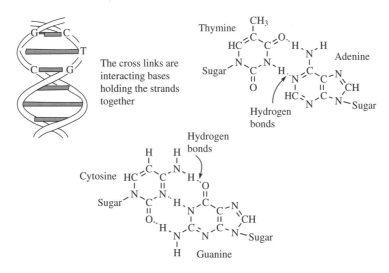

Figure 5.9. Hydrogen bonding in DNA.

Answers to the diagnostic test

1. Methyl amine or amino methane, CH_3NH_2 (2)

2. Amino acids (1)

3. (i) $NH_2 \cdot CH_2 \cdot COOH + HCl \rightarrow (NH_3^+CH_2 \cdot COOH)Cl^-$ (2)

   (ii) $NH_2 \cdot CH_2 \cdot COOH + NaOH \rightarrow NH_2 \cdot CH_2 \cdot COO^-Na^+$ (2)

4. $NH_2 \cdot CH_2 \cdot COOH + NH_2 \cdot CH_2 \cdot COOH$
   $\rightarrow NH_2 \cdot CH_2 \cdot CO \cdot NH \cdot CH_2 \cdot COOH + H_2O$ (2)

5. The $CO \cdot NH$ bond is broken open (1)

Further questions

1. What are the differences and similarities between an amine and ammonia?

2. How similar are amines and amino acids?

3. Use the molecule $NH_2 \cdot CH_2 \cdot COOH$ to show the different forms of this amino acid in acid and alkaline solutions. What is the general name for molecules that have charges at either end of the molecule (ion)? Explain the mechanism by which this molecule could act as a pH controller or 'buffer'.

4. How is the amino acid $NH_2 \cdot CH_2 \cdot COOH$ converted into a tri-peptide? Show its structure.

5. What are the names of the fundamental building units used to make up RNA and DNA?

6. Hydrogen bonding occurs in the DNA helix. Between what elements does this occur? What function does it perform?

7. List some ways in which proteins can lose their 'active' powers and become 'denatured'.

References

1. The special issue of *New Scientist* of 15 March 2003 to mark the 50 years since the discovery of the structure of DNA.
2. The master protein. *Education in Chemistry*, 2002, **39**(6, Infochem), 2–3.

6 Vitamins, Steroids, Hormones and Enzymes

Learning objectives

This unit brings together some of the basic chemistry looked at in earlier chapters and applies it to some important biochemical molecules:

- To show the properties of selected chemical groups and their application to vital molecules used by the human body.

- To show how the body depends on vital chemicals.

- To apply chemical knowledge to show the need for a balanced diet for healthy living.

Diagnostic test

Try this short test. If you score more than 80 % you can use the chapter as a revision of your knowledge. If you score less than 80 % you probably need to work through the text and test yourself again at the end using the same test. If you still score less than 80 % then come back to the chapter after a few days and read again.

1. What is a vitamin? (1)

2. Vitamin C mainly occurs in which foods? (1)

Chemistry: An Introduction for Medical and Health Sciences, A. Jones
© 2005 John Wiley & Sons, Ltd

3. For what is cholesterol best known?   (1)

4. Enzymes do what to chemical reactions occurring in body cells?   (1)

5. Testosterone controls the rate of growth of muscles and which organs?   (1)

Total 5 (80 % = 4)
Answers at the end of the chapter.

My sister is pregnant and the doctor has given her vitamin folic acid tablets to take. My grandmother is 84 and very absent-minded and the doctor has also given her folic acid tablets. She is not pregnant. My father, who recently had a mild heart attack, was also prescribed some foliate tablets. Why has the doctor given all three similar tablets?

6.1   Vitamins

'I take my vitamins and train in gym,

Don't eat junk food and drink no gin.

Keep off fats and eat my greens,

When outside put on sun screens.

Do my exercises and run quite stealthy

I do all this because I want to die healthy.'

There has, in recent years, been some debate about the need to take vitamin tablets. Vitamin is a compound word derived from 'vital amines'. The vitamins are a range of essential chemicals needed by our body. They were given letters of the alphabet in the order they were discovered and the letters bear no relationship to their structures or order of importance. It was originally thought that there was only one 'B' vitamin, but when it was eventually analysed it was realized that it was a complex mixture of a number of compounds, hence the various numberings of the B vitamins. Some of the materials thought to be separate vitamins were found on purification and accurate analysis to be identical to others already given a letter. These were then eliminated. Hence there is no vitamin F or G.

Vitamins are used in very small quantities by our bodies and help to make other materials needed for healthy living. Vitamins are of two types:

- those that are fat-soluble or water-hating (hydrophobic) (A, D, E and K);

- those that are water-soluble (B and C).

The water-soluble ones have polar, water-loving (hydrophilic) groups in their structures, e.g. O—H groups.

The fat-soluble vitamins can dissolve in the fats of our bodies, in which they can be stored. Too large a concentration of these can lead to an illness called hypervitaminosis. Seal livers, for example, are extremely rich in vitamin E. This makes them toxic if eaten in too large a quantity by most people. The metabolism of the Inuit people, who eat seal liver, has adapted to cope with this. The water-soluble vitamins cannot be retained in the body for long periods. They are flushed out and excreted and so must be constantly taken in by eating fruit and fresh vegetables.

'Grandma Dushi has developed a few patches of discoloured skin on her face and arms. I hope it is not catching!' In more recent times Britain, being in northern latitudes, has had an influx of residents from more sunny areas such as the Caribbean and Asia. In some instances the colour of their skin (developed to protect them from very hot sunshine) has prevented them absorbing enough sun through the skin to effectively make enough vitamin D. In these cases the patients are given diets and supplements to try to overcome this deficiency and the resulting skin discolouration.

6.1.1 Vitamin A (retinol)

Vitamin A (Figure 6.1) is found in milk products and eggs, fish oils and vegetables, particularly carrots. It is needed for stimulating healthy growth and for good vision,

Figure 6.1   Vitamin A

so any gross vitamin A deficiencies can cause stunted growth, night blindness and in extreme cases even blindness. Xerophthalmia (a degenerative drying of the cornea) can be caused by a vitamin A deficiency. It is a vital part of rhodopsin, the visual pigment of the eye.

Another name for vitamin A is retinol, and its chemical structure contains a molecule with many unsaturated $C=C$ bonds and also a CH_2OH alcohol group. It is a fat-soluble vitamin. Large doses of vitamin A have been linked to some birth defects so many pregnant women are advised to avoid food supplements rich in vitamin A.

Cod liver oil, containing vitamin A, was often given to young children to supplement their diets. This was particularly true in the UK during World War II, when food was in short supply. I can still remember the bribes I demanded for taking such a revolting liquid with its evil smell and taste. I still shudder at the thought of it even as I write this! Cod liver oil is now recommended for people with arthritis as it is incorporated into the cartilage cells and helps prevent their break-down.[1] This property depends more on the fatty acids present than vitamin A.

6.1.2   The vitamin B series

There are a number of vitamins associated with the letter B. They are all water-soluble so they need to be constantly taken in by eating a balanced diet. They do not have any common structural features. Each has a slightly different role as a co-enzyme (or helper) in regulating and promoting energy release from foods, leading to healthy skin, muscles and general health.[2]

Vitamin B_1, thiamine (Figure 6.2), helps carbohydrates to release energy; it also plays a part in making substances that are nerve response regulators. A major B_1

Figure 6.2   Thiamine, vitamin B_1

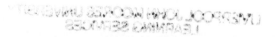

deficiency can lead to 'beriberi', a disease that causes muscle and heart weaknesses. Many foods contain B_1, particularly liver, kidney, pork, yeast, eggs, green vegetables, cereals and fresh fruit. Some foods, like breakfast cereal, have thiamine added to them to ensure a sufficient daily supply.

Vitamin B_2, or riboflavin (Figure 6.3), is a co-enzyme, or helper, needed in the break-down of fats, carbohydrates and especially the proteins needed by the respiratory system. It also helps to synthesize essential mucous material. A deficiency can lead to cracking of the lips (chellosis). It is found in meats, fish, egg yolks, liver, whole grains and dark green leafy vegetables. Remember Popeye in the cartoons? So eat spinach to give you strength! Popeye didn't have cracked lips, but Bluto probably did when he was knocked out by Popeye, often because he had his eye on Olive Oyl, Popeye's girlfriend!

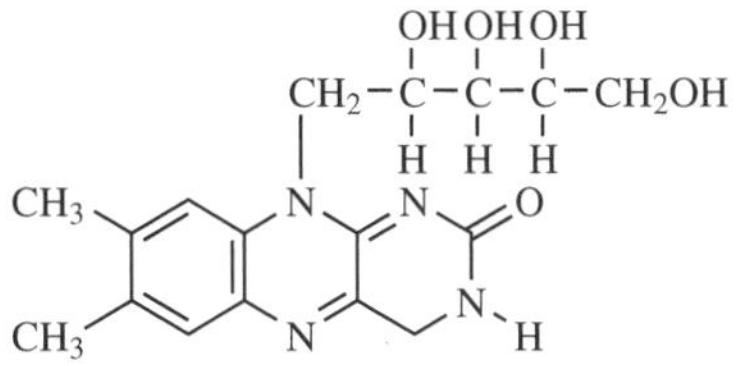

Figure 6.3 Vitamin B_2, riboflavin

Vitamin B_3 (Figure 6.4) releases energy from food. A lack of it can lead to 'pellagra', which looks like skin blisters when exposed to sunlight. In extreme cases it can affect the central nervous system and lead to depression and mental disturbances. Meat and liver products, nuts, vegetables, cereals, salmon and tuna all contain B_3. In larger amounts it has been shown to reduce the cholesterol in the blood, although care has to be taken not to overdose since excess can cause liver damage.

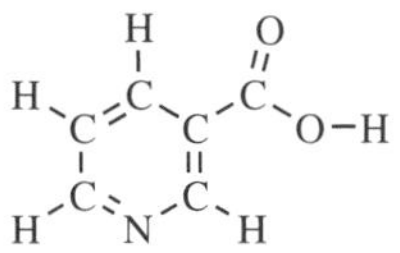

Figure 6.4 B_3, niacin, $C_6NH_5O_2$ or nicotinic acid

Vitamin B_5 or pantothenic acid, $C_9H_{17}NO_5$, is involved in the break-down of fats and fatty acids. It is produced by intestinal bacteria in humans. Its deficiency is virtually unknown.

Vitamin B_6 or pyridoxine, $C_8H_{11}NO_3$, is required in the uptake and metabolism of amino acids, fats and formation of red blood cells. A balanced meat, fruit and cereal diet produces enough B_6 for healthy living. Deficiencies can lead to cracked skin at the corners of the mouth, dizziness, nausea, anaemia and kidney stones.

Vitamin B_{12}, cobalamin or cyanocobaltamin, $C_{63}H_{88}CoN_{14}O_{14}P$, has the most complicated structure of the B vitamins. It contains one atom of cobalt in its molecule. It is required in minute amounts but without it the manufacture of proteins and red blood cells is affected. A diet which is deficient in liver, eggs, meat, fish or milk can lead to pernicious anaemia. Vegetarians are often advised to take vitamin B_{12} supplements.

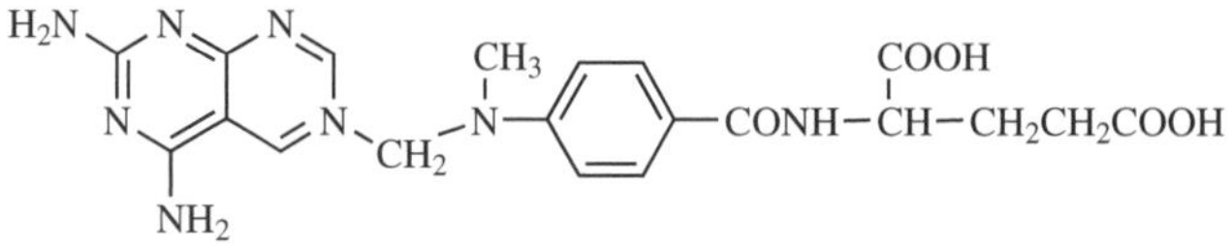

Figure 6.5   Folic acid

Other B vitamins include folic acid, B_9, $C_{19}H_{19}N_7O_6$ (Figure 6.5), which is needed for effective formation of haemoglobin. Although it is found in a balanced diet, any food that has been stored or frozen loses its folic acid and so fresh foods are preferable. Pregnant women are often given folic acid supplements to help in make haemoglobin for themselves and their unborn child. Folic acid is quite a simple molecule that has been known for a number of years to be helpful in ensuring a healthy life and preventing heart attacks and some cancers. It is a water-soluble vitamin so it needs to be taken in constantly. It is present either as the acid or as one of its salts in green dark vegetables, fresh fruit, liver, Marmite and breakfast cereals. However, it has recently come to light exactly what is affected by the absence or presence of folic acid. Any deficiencies of foliates can damage DNA synthesis and this can lead to many cell complications, including the formation of cancers.

Folic acid supplements seems to protect women from giving birth to babies with certain problems, including neural tube defects (NTD). NTD can cause the incomplete development of brain and spinal chord or the protective coverings around these organs. Spina bifida is an NTD defect and affects one baby in 1000. This condition is caused when the spinal chord fails to become enclosed during the early stages of pregnancy. Expectant mothers are now routinely given folic acid supplements and also women wanting to become pregnant. However, as many as 50 % of pregnancies are unexpected or unplanned and so no pre-doses are possible and doses after the first month of pregnancy can be too late to prevent NTD.

Recent research has also noted that intake of foliates can reduce the occurrence of strokes and heart disease in middle-aged and older people. The action of the foliates in the blood reduces homocysteine levels (an amino acid). High levels of this are thought to damage coronary arteries by making harmful free radicals. These free radicals damage cell walls and cause the arteries to fur up, encouraging the blood-clotting platelet cells to gather together and cause a clot.

Low foliate levels and high homocysteine levels are found in the blood of older people, particularly those with Alzheimer's disease. So here is another reason to have a healthy balanced, fruit/vegetable diet throughout your life. There are other B-type vitamins present in minute quantities whose exact role has yet to be fully understood.

6.1.3 Vitamin C

Vitamin C (Figure 6.6) is water-soluble and quite a small molecule. A lack of vitamin C can lead to a complaint called 'scurvy'. In the days of the sailing ships and long sea voyages, British sailors knew about this and on long journeys, say to Australia, would take fruit on board, particularly lemons and limes. They were called 'Limeys' by the Australians, which is still used as a nick-name.

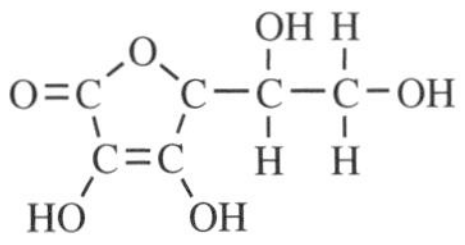

$$O=C \underset{\underset{HO}{C=C}}{\overset{O}{\diagup}} C-\overset{\overset{OH}{|}}{\underset{\underset{H}{|}}{C}}-\overset{\overset{H}{|}}{\underset{\underset{H}{|}}{C}}-OH$$

Figure 6.6 Vitamin C or ascorbic acid

A constant supply of vitamin C is essential for healthy skin and everyday healthy living. 'An apple a day keeps the doctor away' is an old wives' tale, but it has some meaning when applied to vitamins C and B.

Vitamin C occurs in all fruit, citrus, green vegetables, tomatoes, peppers, sprouts, broccoli, etc. This water-soluble vitamin is easily lost from the body and so needs constant replacement by eating fresh vegetables and fruit. On the other hand, excessive intake of it can adversely affect the working of vitamin B_{12} and also cause a loss of calcium from the bones. Excessive vitamin supplements are not usually required if a balanced diet is eaten.

Vitamin C has been shown to have a good effect in destroying the build-up of nitrosamine compounds in the body, which, if left in the body too long, can produce cells leading to tumours. Research reported in 1999 at the American Chemical Society, meetings stated that 'consuming a very large amount of vitamin C may be an effective way of relieving stress'. It must be said that the research was mainly conducted on a rat population, but the research is continuing to see if it has any application to humans. So if you know of any stressed out rats, then feed them fruit!

6.1.4 The vitamin D series

There are two closely related types of vitamin D, D-2 and D-3. Both are fat-soluble and therefore can be stored in the body. Vitamin D_2 (calciferol) can be taken in by eating any balanced diet including vegetables and milk products. Vitamin D_3 (Figure 6.7) is produced in the skin by biological processes that are best catalysed

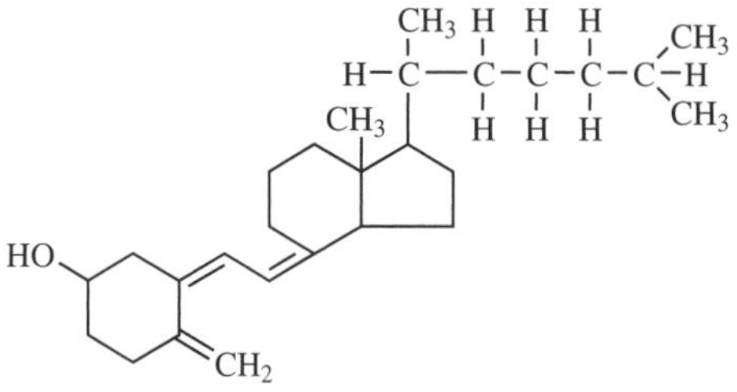

Figure 6.7 D_3, cholecalciferol

by sunlight. Sunlight helps to metabolize calcium–cholecalciferol in the skin into usable vitamin D. We need sunlight to produce enough vitamin D to keep us going through the UK winter, with its low light intensities (even worse in the polar regions). Low vitamin D concentrations can lead to bone disorders including bone loss, osteoporosis and rickets. The health service does give guidelines on how much D is needed. It suggests vitamin supplements to various people, including the elderly, who often stay indoors or cover up from sunlight. Approximately 10 mg of vitamin D is required each day. Even a small amount of sunlight is beneficial, but it must be of sufficient duration or intensity to be fully effective. In tropical zones the sunlight is sufficient to be absorbed through thin clothing. In northern climates it is probably best to absorb sunlight directly. In more recent times Britain, being in northern latitudes, has had an influx of residents from sunnier areas such as the Caribbean and Asia. In some instances the colour of their skin (developed to protect them from very hot sunshine) has prevented them absorbing enough sun to effectively make enough vitamin D. Some people have skin discolouration as a result of lack of vitamin D and a supplement might be suggested.

Some hip bone deformities can be due to this problem and in some communities it is noticed that older Asian women now living in the UK have a pronounced 'roll' in their walking and movement due to this deformity. For such communities a vitamin supplement is often suggested.

Vitamin D is necessary to allow the body to metabolize calcium and phosphorus effectively. Independent of the formation of vitamin D, it must be emphasized that excessive uptake of UV sunlight can lead to abnormal skin cell division and the possibility of skin cancer.

6.1.5 Other vitamins

Other vitamins are needed in small quantities, but all of them are acquired in sufficient quantities by eating a balanced diet. They include:

- Vitamin E, (tocopherols) $C_{29}H_{50}O_2$ (Figure 6.8), which acts as an anti-oxidant to prevent the free radical oxidation of lipids in cell membranes. There are sufficient supplies of vitamin E in vegetable oils.

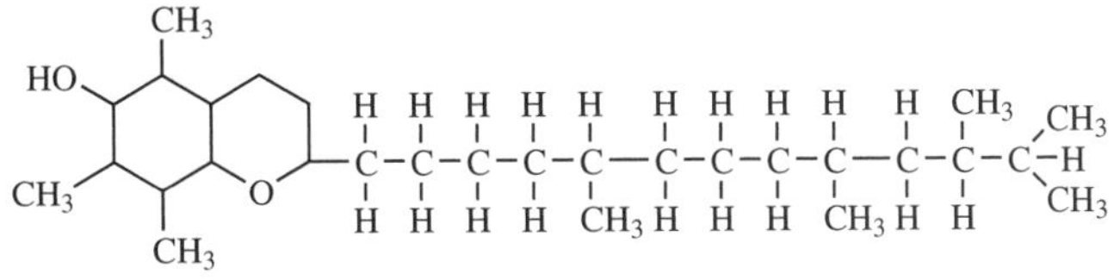

Figure 6.8 Vitamin E

- Vitamin H, $C_{10}H_{16}N_2O_3S$, or biotin (Figure 6.9); deficiency of this vitamin is rare as most food stuffs contain biotin.

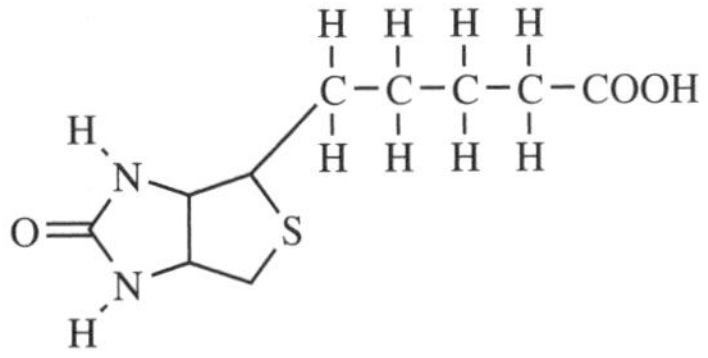

Figure 6.9 Vitamin H, biotin

- Vitamin K, menadione, $C_{11}H_8O_2$ (Figure 6.10), which is found in spinach and cabbage and a wide range of other foods and acts as a blood clotter. Warfrin, sometimes administered as a 'blood thinner', works by disrupting the activity of vitamin K. Vitamin K is also mostly produced by intestinal bacteria. There are some variations of the structure for four similar vitamin K molecules all based upon the structure shown.

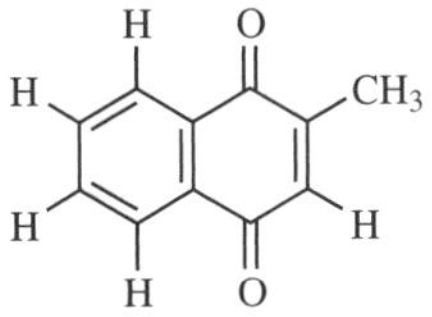

Figure 6.10 Vitamin K, menadione

Have you noticed how most of these vitamin deficiencies can be overcome by a fully balanced diet? Some young people have hang-ups over eating meat and vegetables, depriving themselves of healthy lives in later years. We will leave you to work out what to eat and how best to advise patients and young people.

6.2 Steroids and hormones

At the Seoul Olympics 1988 the Canadian athlete Ben Johnson was stripped of his gold medal and the title of the fastest runner ever for taking a banned substance, an anabolic steroid. So what are steroids and why are they banned from athletics?

Dwain Chambers, the European 100 metres sprint champion, was found guilty in February 2004 of taking a 'designer steroid', THG, and banned from athletics for 2 years. He appealed on the grounds that THG was chemically related to banned substances but was itself not a material that constituted a doping offence.

All steroids contain the basic structures in Figure 6.11. Steroids are fat-soluble lipids. Their structures are based on the tetracyclic (four rings) structure shown in Figure 6.11. Three of the rings have six carbons in them, whereas the other have five

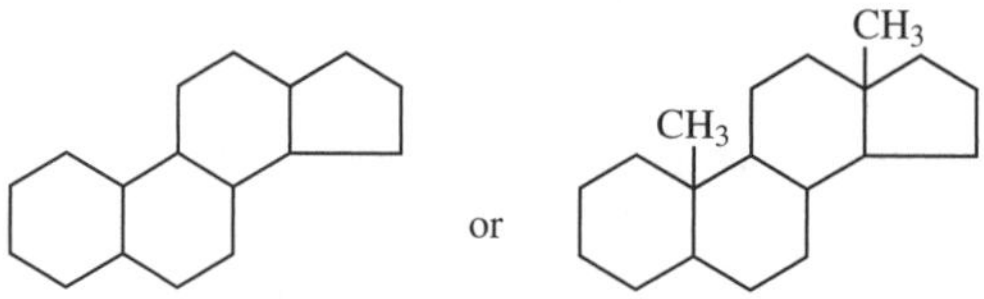

Figure 6.11   General structure of steroids

carbons. Essential compounds like cholesterol and the sex hormones progesterone and testosterone are all steroids. Other examples of steroids are hydrocortisone (a hormone) and digitoxigenin, isolated from foxglove flowers, which is used as a heart stimulant. Some of the old wives' tales about treating heart complaints often used foxglove along with other locally grown plants.

Cholesterol (Figure 6.12) is a very common starting point for making many interesting compounds essential to our bodies.[3] These include bile acids (which help dissolve cholesterol in ingested food), steroid hormones and vitamin D (note some

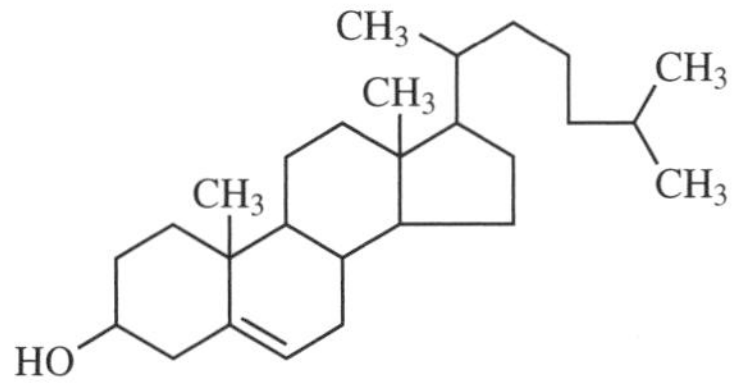

Figure 6.12   Cholesterol

similarities in structures between the compounds listed). The adverse effect of cholesterol is to block arteries, particularly those of the heart. This causes heart attacks. Chemical research is being conducted to try to find chemicals that will prevent this and yet not affect other metabolic processes. A whole range of possible drugs include compounds called 'statins', e.g. fluvastatin. The statin drugs act by inhibiting the enzyme that is involved in the synthesis of cholesterol in the liver. The liver then meets its cholesterol requirements by extracting it from the blood, so lowering the blood cholesterol. Even when eating a no- or low-cholesterol diet, the body manages to synthesize about 800 mg of it each day. It is a required chemical for making other vital materials.

Testosterone (Figure 6.13) is a chemical that controls the rate of growth of reproductive organs, hair growth and muscles. Its presence gives the deepening of the voice in males.

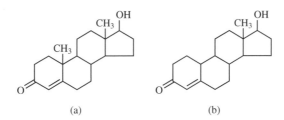

(a)   (b)

Figure 6.13   (a) Testosterone and (b) nandrolone

Some steroids enhance physical performance by increasing muscle growth at puberty. These are called anabolic steroids and are a banned drug for athletes because they create abnormal growth of muscles. Steroids can also help damaged muscles to heal faster. In the Seoul Olympics in 1988, Ben Johnson was found to have illegally taken an anabolic steroid to help him win a gold medal and run 100 m in 9.79 s. He was stripped of the medal as it is a prohibited drug for sports people.

A similar drug, nandrolone (Figure 6.13), has been detected in unusually large quantities in some sportsmen and women and they have been banned from their sports. There is some debate as to whether the body naturally makes nandrolone from testosterone or related compounds. Food supplements given to some athletes by their trainers increase the quantities slightly above the values found acceptable (2 ng/ml sample). Some food supplements show less than 0.6 ng/ml. Greg Rusedeski was accused in January 2004 of being above the accepted level from a sample taken in 2003. He disputed the findings as he maintained he had not taken any unlawful product. He was restored to the professional tennis ranks after investigation.

Although steroids can enhance performance they also can have a long-term effect of causing liver damage, hepatitis and possibly cancer.[3]

Progesterone (Figure 6.14) and a similar compound, oestrogen, are female sex hormones. They cause the periodic changes in the uterus and ovaries during the menstrual cycle. A high level of progesterone is maintained during pregnancy and this prevents ovulation. Hence some birth control pills have focused upon this property

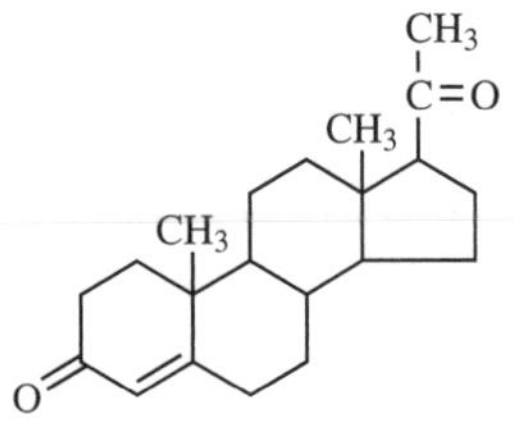

Figure 6.14   Progesterone

Herbalists often use the old and effective recipes gathered from books or handed down through common usage. One such plant is red clover, which contains four different plant isoflavones. When isolated and consumed in a suitable form, these metabolize inside the female body and produce hormone-like compounds which can help with some of the symptoms of the menopause. Research is being conducted to see if these extracts are suitable replacements for the slightly detrimental HRT products. It has been reported that in countries like Japan, where the leguminosae family of plants (lentils, soy beans, sprouted beans, ground nuts) are consumed as foods, menopausal complaints are relatively rare. This has been put down to the phytoestrogens or plant hormones found in these foods.[4]

6.3   Enzymes

Enzymes are usually large long-chain protein molecules that act as a catalyst for chemical reactions in the body. A catalyst is a material that speeds up a chemical reaction but is itself unchanged at the end of the reaction. Very small amounts of catalysts and enzymes can affect large quantities of reactants. They can, however, be poisoned if the conditions are changed and become unfavourable, or an impurity is present. Enzymes, being proteins, like to work at a set and narrow range of temperatures, and if they are warmed above their working temperature they become inactive due to being de-natured.

Enzymes cannot perform reactions that would not normally take place, they simply speed them up. One of the key features of enzymes is their characteristic and specific shape. One enzyme does not work with another reaction because the reactant and enzyme interlock and this depends upon their mutual matching shapes. Enzymes are usually specific to one type of reaction.

Catalase is one of the vital enzymes carried around in very small quantities in the blood. It destroys the harmful effects of any peroxide molecules or ions being produced accidentally when glucose is oxidized. If it was not destroyed this would cause much harm to many of the body's systems, including destroying the DNA of cells. The peroxide is converted into the harmless materials of water and oxygen gas:

$$2H_2O_2 \rightarrow 2H_2O + O_2$$

Another useful enzyme is amylase, found in the saliva and the small intestine. It has a role in the digestive system, breaking down the starches (carbohydrates) in foods into maltose. This is then acted upon by a further enzyme, maltase, to convert the maltose into glucose. Humans can digest the starch in potatoes but the enzymes do not break down the chemically similar substance, cellulose, into glucose. Thus humans cannot digest the cellulose in grass, but the enzymes in the digestive systems of cows and sheep breaks it down.

The human stomach and small intestine also contain enzymes that help in the hydrolysis and break-down of proteins, first into shorter chain peptides (this is done with the aid of the enzymes pepsin and trypsin), and then hydrolysed further into individual amino acids with the help of the enzyme peptidase. Any fats in food are also hydrolysed in the stomach with the aid of lipase enzyme to form fatty acids (carboxylic acids).

It can be seen that the break-down of the components of food is greatly speeded up with the aid of enzymes, which make the smaller units of glucose, amino acids and fatty acids available for transport to cells for synthesis into body-building materials.

6.3.1 How enzymes work

There have been many theories of the exact mechanism of the action of an enzyme because so little of it is needed to affect the rate of a reaction. Some enzymes can speed up a reaction by as much as 10–20 000 times. One explanation for the mechanism is as follows.

In a normal reaction the two reactants need to collide with each other many times for a reaction to occur. In each collision only a few of the molecules are in the correct orientation to actually react with their counterparts, so owing to the randomness of motion, the reaction can take some time. When an enzyme is used, the reactant fits onto the enzyme molecule in the correct specific orientation so that, when the other reactant comes close to the enzyme/reactant combination, it is immediately in the correct orientation and alignment for the two reactants to join up. These react and then break away from the enzyme and form the new molecule, and the enzyme is available for reuse. Because the reactants are lined up correctly, no time is lost in useless random collisions between the reacting molecules. This mechanism is sometimes referred to as the 'lock and key' model (Figure 6.15).

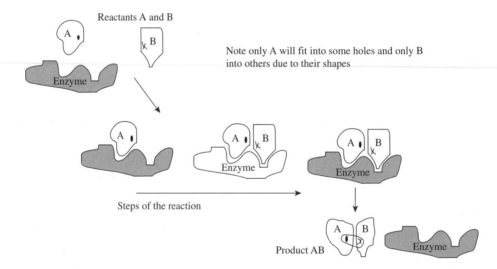

Figure 6.15   Mechanism of the action of enzymes

6.3.2 Applications of enzyme chemistry

- Enzymes are sometimes enhanced and supported in their activity by smaller molecules or metals ions. These reactants are called co-enzymes. We will not look further into the mechanisms of co-enzymes.

- Enzymes are not only used in our bodies to speed up reactions; they are also commercially used. One such case is 'enzyme'-enriched detergents, which 'eat away the dirt'.

- There are many other clever applications of enzymes in the chemical industry. For example, the soft centres of some hard boiled sweets contain a minute quantity of an enzyme. It slowly reacts to soften the sweet from the centre outwards and so causes the centre of the sweet to be soft and chewy.

- Enzymes in yeast help ferment sugars to make alcohol (ethanol).

- Enzymes can be poisoned if an impurity molecule blocks up the active site of the enzyme. Some enzymes can be used in treatments to prevent unwanted reactions occurring, like the production of bacteria.

- Some diseases are caused by enzyme deficiencies. The congenital disease phenyl-ketonuria is caused by the deficiency of the enzyme phenylalanine hydroxylase, resulting in a build-up of compounds that cause brain damage and mental retardation. This damage can be lessened and prevented by a diet containing a low amount of the amino acid phenylalanine. The complaint is caused by a genetic mutation.

- Albinism is also caused by lack of the enzyme tyrosinase, resulting from a genetic defect.

- Many heart attacks are caused by a blood clot forming in a coronary artery, and recently this has been found to respond to the enzyme streptokinase to dissolve the clot. It is injected to as close to the clot as possible.

- When blood is required to clot, as in the case of a cut or wound, enzymes are involved. Thrombin catalyses the soluble protein, fibrin, in the blood to form the insoluble fibrin, helping clotting and wound repair.

- Amylase in the saliva and small intestine helps convert starch to maltose, and this in turn is converted into glucose in the small intestine with the enzyme maltase present. Glucose is either used for energy release or stored as glycogen for future use.

- Gastric enzymes such as pepsin and trypsin convert proteins from food into the smaller peptide molecules, which in turn are converted into amino acids.

- In the small intestine there is lipase, which converts fats (or lipids) into fatty acids.

- The disaccharide lactose, is only found in milk and is hydrolysed with water in the presence of an enzyme lactase to form the monosaccharide glucose and galactose. This occurs in the small intestine. Some people, and particularly those from Eastern and African countries, are deficient in lactase so they are intolerant of milk. This can cause diarrhoea. Many African and Chinese foods do not include milk for this reason.

- Antabuse, or disulfiran, is used to deter alcoholics from drinking alcohol again. They take the drug, and then when they consume alcohol they feel sick and nauseous. This is because the antabuse inhibits the enzyme that aids the oxidation of ethanal, an aldehyde which is one of the products of alcohol oxidation in the system. The build-up of ethanal makes the person feel sick. The effect slowly wears off but leaves a memory of nausea.[5]

- It has been known for a long time that the presence of too high a concentration of free radicals can have a damaging effect on the DNA of a cell by its electrons being stolen by the free radical. However, it has only recently been realized that all organisms have enzymes that can repair any damage caused by this process. Exactly how the enzyme travels along the long DNA chains to find the damaged parts and repair them is a mystery waiting to be unravelled.[6]

Answers to the diagnostic test

1. Essential chemicals for body reactions   (1)

2. Fruit and vegetables   (1)

3. Furring up arteries and causing heart attacks   (1)

4. Speed up chemical reactions   (1)

5. Sex organs   (1)

Further questions

1. On the instructions of an enzyme-enriched washing powder it says 'soak the dirty clothes in the water + washing powder overnight, then wash as normal the next day'.

   i. Explain why it is necessary to soak in cold water overnight.
   ii. Why not boil and wash the clothes in the hot water + enzyme powder straight away?
   iii. Will there be any dirt that the enzyme powders will not break down?

2. List three of the important uses the body has for enzymes and also three commercial uses of enzymes.

3. From your knowledge of vitamins what advice would you give to the following groups of people to ensure healthy living:

   i. elderly people;
   ii. pregnant women;
   iii. vegetarians and vegans;
   iv. mothers feeding young children;
   v. teenage girls afraid of putting on weight?

4. People say 'cholesterol is a killer'. What advice can you give about this statement to alleviate anxiety and give them a better understanding of the role of cholesterol in the body?

5. Give an overview of the necessity to encourage young people to have a balanced diet.

References

1. B. Caterson. Fishing for arthritis drugs. *Chemistry in Britain*, April 2000, 20 (summary of *Journal of Biological Chemistry*, 2000, **275**, 721).
2. R. Kingston. Supplementary benefits. *Chemistry in Britain*, July 1999, 29–32; *Chemistry in Britain*, November 1996, 38.
3. S. Cotton. Steroid abuse. *Education in Chemistry*, May 2002, **39**(3), 62.
4. Reported in *Country Life*, May 2000, 134ff.
5. S. Cotton. Antabuse. *Education in Chemistry*, January 2004 **41**(1), 8.
6. A. Ananthaswamy. Enzymes scan DNA using electric pulse. *New Scientist*, 18 October 2003, 10.

7 Ions, Electrolytes, Metals and Ionic Bonding

Learning objectives

- To understand the general principles of ions and ionic bonding.

- To understand some important functions of both anions and cations in the body.

- To appreciate the diverse uses of elements and ions in various metabolic processes of the body.

Diagnostic test

Try this short test. If you score more than 80% you can use the chapter as a revision of your knowledge. If you score less than 80% you probably need to work through the text and test yourself again at the end using the same test. If you still score less than 80% then come back to the chapter after a few days and read again.

1. If an element is in group 1 of the periodic table how many electrons will it have in its outermost shell? (1)

Chemistry: An Introduction for Medical and Health Sciences, A. Jones
© 2005 John Wiley & Sons, Ltd

2. What happens to the electrons of sodium when it reacts with chlorine to form sodium chloride? (1)

3. A solution conducts electricity. Does it contain a covalent or an ionic compound? (1)

4. Do ionic compounds have high or low melting points? (1)

5. A solution containing ions is also called an e—? (1)

6. Where in the human body would the greatest amount of iron be found? (1)

7. A healthy thyroid requires a constant supply of which ion? (1)

8. Approximately 85 % of the phosphorus found in our body is located where? (1)

9. Which two ions constantly pass through the cell walls of our bodies, usually from opposite sides? (2)

Total 10 (80 % = 8)
Answers at the end of the chapter.

'No Gold for Jim', read the big headline in the papers. Jim had led the final stages of the Olympic marathon for at least 10 miles and as he entered the stadium the applause was deafening but with one lap to go the strength in his legs began to fail. He began to roll and with only a hundred yards to go he began to crawl and his body was losing its coordination. By then the next runner had entered the stadium, and the next. The crowd were urging him on but with no effect' and the doctors had to step in and treat him, meaning immediate disqualification. He recovered after a drink and restoration of the electrolytes in his body fluids. He received a tremendous ovation but no Olympic medal.

The movement of sodium and potassium ions back and forth through the cell membranes gives rise to electrical impulses and nerve transmissions within the body. This can become disrupted with increasing dehydration.

Any water-soluble material, including foods and body fluids, contains positively and negatively charged ions. Many of the cell processes are controlled by adjusting the concentration of these ions entering or leaving the cells of our body. Get these concentrations out of balance, and messages around the body go haywire; thinking is affected and cell stability is disturbed.

7.1 Introduction to ionic bonding

You will already realize from earlier chapters that, for an atom to gain stability when it forms a compound, it tries to gain or lose electrons to make its outer electronic structure that of an 'inert gas' (which has a complete outer electron shell). Some atoms in molecules achieve this by 'sharing' electrons, as in the case of covalent bonding. There is another way of achieving outer electron stability. This is by complete donation, to another atom, or complete acceptance of electrons from another atom. In so doing positive or negative ions are created. These two oppositely charged ions attract each other, tightly holding onto each other. The bonding holding these charged particles together is called 'ionic bonding'. A typical example of a material containing ions is sodium chloride, NaCl.

- Sodium, Na, atoms have an electronic structure of 2.8.1;

- Chlorine, Cl, atoms have an electronic structure of 2.8.7.

In the compound sodium chloride, stability is achieved by loss and gain of electrons so that each has a complete outer electron shell (Figure 7.1). The negative ion, Cl^-, is called an anion and the positive ion, Na^+, a cation.

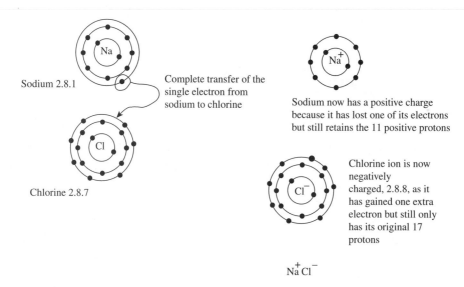

Figure 7.1   Structure of sodium chloride

Sodium chloride, or 'common salt', is water-soluble, as are most other ionic compounds. Because the solution contains a charged set of ions, it will conduct electricity. Such electrically conducting solutions are generally called 'electrolytes'. Electrolytes are solutions containing ions which can freely move around. The ions are attracted to areas or electrodes of opposite charge. They are repelled from areas of the same charge. The freedom of movement of the ions means that the solution conducts electricity.

7.2   Some common properties of ions and ionic bonds

- Ions come from ionic compounds; they have stable structures, with their outer electron shells complete. They can be positively charged, as is the case with all metallic ions. The number of charges on a metal ion corresponds to the group of the periodic table it is in, e.g. sodium, Na^+, is in group 1, calcium Ca^{2+}, is in group 2 etc. Negative ions are generally composed of nonmetals or groups of them joined together. Nonmetal ions include chlorides, Cl^-, bromides, Br^-, sulfates, SO_4^{2-}, carbonates, CO_3^{2-}, nitrates, NO_3^-, and phosphates, PO_4^{3-}.

- Water is able to prise apart ions in a solid ionic compound and dissolve them. Some ionic compounds are more soluble than others.

- Aqueous solutions of ionic compounds will conduct electricity if positive and negative electrodes are connected to a DC source and inserted into the solution (Figure 7.2). The positive metal ions, cations, slowly migrate to the negative electrode (cathode) and the negative ions, anions, migrate to the positive electrode (anode).

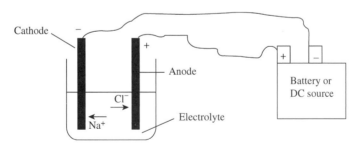

Figure 7.2   Electrolysis

- In the solid state, ionic compounds are held together strongly by their positive and negative charges, so the compounds have high melting and boiling points. NaCl melts just above $800\,°C$.

- The hundreds of ions holding each other together make ionic compounds giant structures, but the ratio of the positive to negative ions is always the same, e.g. one Na^+ to every one Cl^- (Figure 7.3).

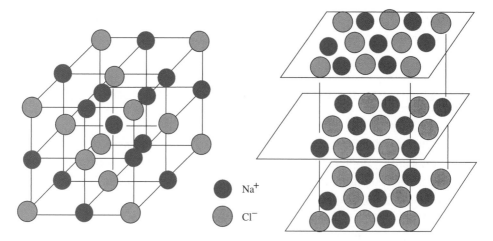

Figure 7.3 Sodium chloride crystals. The lines are drawn only to show the orientation one to another and do not imply that the bonds in ionic compounds are directional or stick-like

- Ionic compounds do not have a smell because the charged particles are held together strongly and do not allow any of their particles to escape into a vapour (unlike covalent compounds, whose intermolecular forces are weaker, so some of their molecules escape as a vapour, hence they smell).

- Because the ions are strongly held around each other, they attract each other in all directions over the surface of the ions. Thus there are no characteristic bond angles in ionic compounds (cf. the specific bond angles in a covalent molecule). Ionic crystals are large collections of ions held together in an ordered pattern to form characteristic crystal shapes. Sodium chloride is a cubic crystal, other compounds also have shapes such as pyramids or distorted cubes, depending upon the size of the individual ions.

- Different ions in solution have different effects. Hydrogen ions, H^+, are the cause of acidic properties and hydroxyl ions, OH^-, cause basic or alkaline properties.

- The smaller ions can pass through cell membranes walls, e.g. H^+, Na^+ and K^+.

- Many body fluids are not simple solutions of one type of ion, but are complex mixtures that change composition depending on where they are going and where they have come from.

7.3 Electrolytes and ions of the body

Electrolytes are solutions of positively and negatively charged ions and have several different roles within our bodies:

- They maintain and transport essential ions around the body in the various fluids. They occur within cells, e.g. blood, urine, plasma, and in the fluids between the cells (interstitial fluids).

- They carry essential minerals to where they are required.

- Their concentrations in solution control osmotic pressures in various cells and within various organs.

- The passage of the smaller sodium and potassium ions through cell membranes causes the body to react or stop acting and pass on information or not.

- They maintain the pH control for all systems. This is particularly true for the amino acid and protein molecules or zwitter ions which are good 'buffers'.

- The ions can move in solutions and conduct small electrical currents within cells and between cells of the body. These often control the release of hormones and maintain the effectiveness of neurotransmitters in the brain. A neuron (nerve cell) message is passed by a flow of Na^+ ions into a neuron and a flow of K^+ ions out, giving about 100 mV potential difference at the cell membrane only. This passes along a neuron at about 120 m/s.

Each of the body fluids differ in their content. The plasma cells contains a number of protein ions (remember amino acids and proteins can form ions and zwitter ions), whereas the fluids between the cells (intercellular fluids) contain fewer protein ions. Plasma moves around inside the blood vessels which are impervious to large protein molecules passing through their membranes into the interstitial fluids.

All cells, including muscle and nerve cells, have inside them an intracellular fluid (ICF) which contains high levels of potassium, K^+, phosphate ions, PO_4^{3+}, and protein and small amounts of Na^+ ions and chlorine ions. Outside the cells in the extracellular fluid (ECF) consists mostly of sodium ions, Na^+, chloride, Cl^-, and bicarbonate ions, HCO_3^-, but no protein, plus low concentrations of potassium ions. The inner layer of the cell membrane is negatively charged relative to the outside. When activity occurs then an ionic pumping action takes place to try to maintain the balance within the cells between the intra and extra flow of sodium and potassium

Table 7.1   Location and purpose of some common ions in the body

Sodium ions, Na^+	Potassium ions, K^+	Calcium ions, Ca^{2+}
90 % of all sodium cations are in fluids outside the cell (ECF)	K^+ is the most abundant cation inside the cell fluids, plasma etc.	98 % of Ca^{2+} is part of bone structures and so is the most abundant ion in the body.
There are fewer Na^+ ions inside cell fluids	It plays a key role in the balance of responses of nerves and muscles	There are also some in extracellular fluids
Essential for conduction of electrical impulses for nerves and muscles	Movement in and out of the cell is balanced with the opposite movement of H^+ ions and so can act as a pH control	
Crucial for keeping the balance between osmotic pressure in extracellular fluids and water		
The Na^+ concentration is controlled by ADH hormone		
Loss of Na^+ by excess perspiration, diarrhoea or sickness results in weakness, dizziness, headache, hypertension etc. (hyponatraemia)	Low K^+ levels (hypokalaemia) results in diarrhoea, vomiting causing cramps and mental confusion, fatigue, high urine output etc.	Diets must contain intake of calcium, otherwise numbness of fingers occurs and bone strength is reduced.
Higher than normal Na^+, hypernatraemia, leads to water moving out of the body cells and gives symptoms of intense thirst, fatigue, agitation and, in extreme cases, coma	K^+ levels can be confirmed by electrocardiogram data	Ca^{2+} ions are essential to the intracellular messenger system
	High K^+ level hyperkalaemia, can cause death by fibrillation of the heart.	Excessive amounts of Ca^{2+} ions can lead to calcification of tissues, cataracts, kidney stones and gall stones
	Na^+/K^+ balance is essential	

ions. Table 7.1 shows some of the location and purposes of the most common ions in our bodies.

7.4   Major cations (positive ions) in the body: sodium, potassium and calcium ions

The cell membranes are made up of protein and fats known as phospholipids and are configured into pores, holes and channels. Some of these channels are able to selectively allow some ions to pass through. Channels exist that allow Na^+ ions to pass and these are called sodium channels. Other pores and channels are selective to allow K^+ or Ca^{2+} ions to pass. These channels are used in the passage of electrical charges and messages along and between nerves cells. They are therefore important in understanding pain killer mechanisms and drug design.

Table 7.2 A summary of other cations present in body fluids. The following elements are present in very small quantities in the body and are ofter referred to as the 'trace elements'. They are present in less than 0.01 % of the body mass

Aluminium ions, Al^{3+}	Thought to have an adverse effect in excessive concentrations. Older research thought it was linked to Alzheimer's disease; this is now less certain. This research is ongoing
Cadmium ions, Cd^{2+}	Exposure to small doses of cadmium metal or ions can increase the risk of cancers, particularly breast cancer. Smoking doubles the normal natural intake of Cd[1]
Chromium ions, Cr^{3+}	Present in very small amounts but essential for metabolism of sugars and blood control. Found in yeast, wines and beer and balanced diets
Cobalt ions, Co^{2+}	Essential for the formation of vitamin B_{12} and synthesis of red blood cells. Without it pernicious anaemia follows. Not found in vegetables but it is present in milk, eggs, cheese, meat and liver. Vegetarians must be aware of their possible cobalt deficiency and remedy this
Copper ions, Cu^{2+}	Approximately 0.9 mg of copper are needed daily. Small quantities are stored in the liver and are needed to help in the synthesis of haemoglobin and also in forming the pigment melanin. A balanced diet including egg, whole wheat, beans, beet, liver, fish etc. is essential. Copper is also found in the brain and eyes. Its compounds are linked to the action of enzymes. It is also needed to make collagen and maintain the body's immune system. Wilson's disease, a genetic metabolic defect, has symptoms of brown or green rings in the cornea of the eyes. It is caused by the accumulation of too much copper in the tissues but it is treatable
Hydrogen ions, H^+	Hydrogen ions occur everywhere in the body in small amounts as an ion but large amounts in water and aqueous liquids. They are also in aqueous solutions and body fluids. The correct balance of H^+ in solution maintains the correct working pH values. pH is buffered frequently by proteins to maintain the best working values. Water is taken in as a liquid and also by eating plant material, etc. It is essential for maintaining H^+ ion concentration
Iron ions, Fe^{2+}, Fe^{3+}	Found in about 5–7 g quantities in human bodies, mainly as the iron in blood haemoglobin (66 %). The other 33 % is found in the liver, skeleton, muscles etc. Red meats, green leaves and root vegetables along with eggs, beans and dried fruit are all rich in iron. Ingested iron cannot be of use for haemoglobin until it has been taken into the bone marrow, where the haemoglobin is constantly synthesized. Iron is crucial for binding oxygen to haemoglobin in the lungs and blood system. The average lifetime of red blood cells is 120 days and 25–30 mg of iron are lost each day, so constant intake of a balanced diet is needed to replace these vital ions, particularly for females. An iron-containing enzyme, tyrosine hydroxylase, catalyses the formation of L-dopa, the rate-limiting step in the biosynthesis of the neurotransmitter dopamine. A deficiency of dopamine production is associated with Parkinson's disease
Magnesium ions, Mg^{2+}	Needed for nerve and muscle tissue and formation of bones. Deficiency is linked with diabetes, blood pressure and pregnancy problems. Plentiful supplies are present in balanced diets containing, grain, green vegetables and sea food. Chlorophyll contains magnesium in its molecules, so eating green plants means you take in magnesium ions. Mg^{2+}–ATP link is an essential co-factor in enzymes that hydrolyse ATP for energy generation

(continued overleaf)

Table 7.2   (*continued*)

Manganese ions, Mn^{2+}	Present in some enzymes, it activates blood haemoglobin synthesis, urea formation, growth, reproduction processes and release of insulin. Manganese ions also take part in enzyme control in the brain
Zinc ions, Zn^{2+}	Small quantities help carbon dioxide metabolism and removal. Zinc is needed for healing of wounds and growth and maintaining healthy sperm count in males. It is present in some enzymes to help protein digestion. A deficiency has been associated with slow learning in children. Zinc ions are also crucial for maintaining enzyme actions in the brain. Deficiencies affect the sensitivity of the sense of smell. High values can increase cholesterol. Zinc is present in small quantities in many foods in a balanced diet

Table 7.3   Anions (negative ions) present in body fluids

Anions present	Properties
Chloride, Cl^-. A small ion able to move through cell membranes	A major extracellular anion, but it easily moves into intracellular places It is important for balancing osmotic pressure In the stomach it becomes linked with H^+ ions to form hydrochloric acid, which is essential for food digestion
Phosphate ions, $H_2PO_4^-$, HPO_4^{2-}, PO_4^{3-}	Approximately 85 % of the phosphate is present in the bones. The rest is in ADP and ATP It is a vital component of the buffer solutions of the body which maintain pH values
Carbonate/bicarbonate ions, CO_3^{2-}, HCO_3^-	It is present in buffer solutions of the body HCO_3^- is a weak acid that helps to maintain pH control within cells These ions are involved in the uptake of CO_2 and carrying it away from the sites where it is produced as a waste gas to the lungs, where it will be exhaled. The ions are also used to regulate the pH within cells and at particular working sites, e.g. muscles

Table 7.4   A summary of other ions or elements present in body fluids

	Location and properties
Carbon is only present as organic molecules and not as single carbon ions	Carbon is present in all living cells as the element in carbohydrates, proteins, lipids, etc. Without carbon we would not exist. Carbon dioxide is exhaled by all animals and used by green plants to synthesize carbohydrates. There is a constant exchange of carbon in living systems
Bromide, Br^- (chloride ions have already been discussed)	In small quantities in the body Br^- can replace Cl^- and cause drowsiness and decline of activity, including sexual drive. Hence the use of the so-called 'bromide pill'. It was said to be added to cups of tea in the forces canteens during World War II – but this was probably only a story!
Fluoride, F^-	Used in foods, water and toothpaste to strengthen the surface materials of teeth – enamel. There is still a debate as to its value

Table 7.4 (*continued*)

	Location and properties
Iodide, I^-	Small quantities are essential for a healthy person to maintain the synthesis of thyroid hormone, which regulates metabolic rates. Present in soil and any food plants, particularly sea foods, as iodide ions
Oxygen, peroxide ions, O_2^{2-}	Oxygen gas is essential for all life as it oxygenates haemoglobin, which then transfers oxygen via the blood throughout the body. Sometimes small amounts of peroxide radicals, O_2^{2-}, are formed. The enzymes peroxidase and catalase in the blood destroy the peroxides as they are harmful to all living tissues
Selenium, Se	Present in very small quantities, it is an anti-oxidant and helps prevent break-down of chromosomes. Its deficiency can cause birth defects. Found in many foods in a balanced diet and also in cereals, mushrooms, garlic, meat and sea foods. It is also a constituent of some anti-dandruff shampoos
Sulfur compounds mainly as sulfate ions, SO_4^{2+}	There are small amounts of sulfur in the body, mainly in a few of the amino acid units present in proteins, whose components are sulfur-containing amino acids. Some of these occur in the proteins in the hair. Some waste body gases contain the break-down products of these sulfur compounds as hydrogen sulfide, smelling of bad eggs

7.5 Balance between fluids

It is essential that there is a balance between intracellular (within cells) and interstitial fluids (between cells) so that they have the same balanced osmotic pressure. Both over- and under-hydration can result in some drastic and serious body and neurological abnormalities due to abnormal flow of the ions between the inter- and intra-cellular fluids.

Two chemists, Peter Agre and Roderick MacKinnon[3] received the Nobel prize in 2003 for their work on ion channels in cells and living tissues. These channels can selectively filter K^+ ions and Na^+ ions from solution. If the transport of these ions cascades, they can cause muscles to contract, cause eyes to water and malfunction of the brain cells and stop the cells' ability to intercommunicate. This could result in cardiovascular problems, liver malfunction, dehydration and other disorders.

The different concentrations of the ion inside and outside the cells cause a slight electrical potential difference across the cell membrane. It is very small, approximately 100 mV. However, any changes in the concentrations of the ions present change this value. Thus concentrations of ions and electrical charges must be synchronized to allow the cell's materials and charges to flow across a membrane. This cycle in the cells of the heart causes the regular pattern of the heart beat and any imbalance of electrical activity or ion imbalance can cause a malfunction of the heart. An artificial pacemaker can be used to maintain the electrical impulses needed by the heart if this system is at fault.

Electrocardiography, can be used in medicine for tracing the electrical activity of the heart. Its rhythmic and orderly beating is maintained by a series of discharges originating in the sinus node of the right atrium. By attaching electrodes, normally to the chest, a small current can be detected. The pattern of the strength of this current over time when graphed out on a recorder is called an electrocardiogram, or ECG. A trained person can analyse the shapes and patterns of the graph and interpret them in terms of normal or abnormal heart activity. The rhythmic pattern clearly shows that the mechanical pumping of the heart is maintained by a systematic flow of ions and an electrical flow within the heart muscles (Figure 7.4).

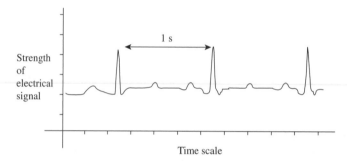

Figure 7.4   Typical ECG pattern monitored by elecrodes attached to a person's chest

7.6   Essential elements present in small quantities: micronutrients and minerals

The metabolism of the body is generally controlled by hormones in conjunction with vitamins, minerals, enzymes and other things, including ions. Minerals are inorganic compounds (i.e. made from elements other than carbon) or their ions and make up about 3–4 % of the body mass. These are mainly found in the skeleton and body fluids.

It has only been in recent years, with an increase in the accuracy of analytical chemistry, that the very small quantities of some of these elements have come to light in the body and have been seen to have vital roles. Whoever would have thought that minute quantities of manganese ions were vital for the synthesis of haemoglobin, bone construction and a healthy sex life? Molydebnum has recently been detected in the body fluids and appears to have some role involving enzymes, but its exact mechanisms are as yet unknown. Most of the elements that are present in very small quantities have the purpose of regulating body functions. It is not quite as simple as 'suck a bar of zinc metal to help your sex life'. Some elements are in fact toxic in too large a quantity and others must be digested in a specific ionic form before the body can use it. Some example of elements or ions that are not usually present in a healthy person but can enter due to uptake from the environmental, foods or direct poisoning are listed in Table 7.5.

Table 7.5   Some harmful metal ions

Lead ions, Pb^{2+}	These ions and the metal are accidentally taken into the body by breathing air polluted by burning leaded petrol, which is now banned. Its presence is long-lasting as it is not easily metabolized. It has been shown to slow down thinking and reasoning processes. Large quantities can lead to death. The banning of leaded petrol and lead-based paints has reduced occurrences
Thallium, atomic no. 81	This element is better known for its poisonous properties and links with almost undetected deaths and 'Agatha Christie'-type murder mysteries.[2] Thallium compounds are used in some countries as cheap insecticides, particularly for killing cockroaches. Intake of this element causes stomach pains, vomiting and nausea, painful soles and palms of hands, limb weakness, double vision, involuntary eye movements, hallucinations, characteristic hair loss and white lines across the nails. The symptoms are often mis-diagnosed as other diseases. Treatment is with the chemical Prussian Blue
Mercury	Mercury salts if taken into our food chain causes sickness, headaches and dizziness. In larger quantities they can lead to severe illness and death. Such a case of mercury poisoning was caused in Japan by people eating raw fish that had bred in the outflow of a factory that was illegally pouring mercury compounds into the sea. The mercury formed methyl mercury compounds with microorganisms which were ingested by the fish. These organic mercury compounds dissolve more easily in fatty compounds of the body of fish and eventually humans. Mercury alloys (mixtures of mercury, tin and silver) are safely used in dentistry. The mercury metal used in some thermometers, if swallowed accidentally by breaking the bulb of the thermometer, can usually pass through the body without causing too much harm

7.7   Cancer treatments and chemotherapies that use metal compounds

Cancer cells are formed when some of the cells are genetically damaged, which prevents them from being controlled by the normal cell mechanisms. If these cells multiply, the abnormalities can be transferred to new cells and these can lead to a rapidly forming tumour. When these tumours are detected there are a number of treatments possible, from its surgical removal to the use of specific chemicals *in situ*. Tumours can also be 'zapped' with accurate doses of radiation. These paragraphs will concentrate on the use of chemicals or chemotherapy.

The search for these chemical combatants often turns to nature's remedies using plant, microbes and fungi, but more recently specific drugs have been chemically synthesised.[4] One platinum compound, cisplatin (Figure 7.5), is widely used for

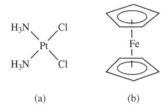

(a)   (b)

Figure 7.5   (a) Cisplatin and (b) ferrocene

treating testicular, ovarian, bladder and neck cancers. It is administered intravenously and from the blood it diffuses into the cytoplasm of cells. Here it forms compounds with the nitrogen on guanine groups on adjacent strands of the DNA. This inside cells distorts its shape and structure and prevents such cells reproducing. Unfortunately cisplatin is not selective to just cancer cells and can attack 'normal' proteins and this can lead to severe side effects, including renal impairment and also loss of balance and hearing, which are often only partially reversed after treatment.

The search for other, and less toxic, water-soluble and more specific drugs has led to a group of compounds known as 'sandwich compounds' being investigated. These compounds have a metal atom held between suitable plate-like molecules. These, under suitable conditions, can be attached to DNA chains, distorting their structure and disrupting the growth of the cancer cells. There is evidence that some of these compounds are better than cisplatin for some cancers but less effective for others. It is possible that a method of combining suitable drugs in a 'cocktail' could be effective so that what one drug misses the other one will get. The mixtures must be carefully formulated and possibly made up individually for each patient. This is called 'combination therapy'.[5]

In order to prevent the active anticancer compounds killing off other fast growing cells, like the hair, methods are being developed to pinpoint the release of active drugs like cisplatin at the exact point of the cancer and nowhere else. One method is to attach the cisplatin to another compound so that the new compound is harmless to healthy cells. The cisplatin can be released from the compound by shining light upon the cancer area, pinpointing the specific cancer cells.[6] Other compounds could be released by change of pH or acidity.

Active research continues, but from the moment of discovery to the point of actual use in human patients can take many years. One method being researched is to develop a material that cuts off the blood supply to cancer cells only. This would starve the cells and prevent growth.

Gold compounds, including auranofin are used in the treatment of rheumatoid arthritis.

Answers to the diagnostic test

1. One (1)

2. They are completely transferred to the chlorine atom (1)

3. Ionic compound (1)

4. High (1)

5. Electrolyte (1)

6. Blood (1)

7. Iodine (1)

8. Bones (1)

9 Sodium and potassium ions (2)

Further questions

1. Give an account of the importance of a balanced diet in providing essential elements and ions for healthy living.

2. 'It is essential to maintain the balance between Na^+ and K^+ ions in cells and surrounding fluids.' Discuss.

3. Ions are present dissolved in various fluids of the body, and some are present in very small quantities. Discuss the importance to human metabolism of small quantities of the following: zinc, manganese, cobalt and iron.

4. Explain what is meant by the 'electrolytes' needed by the body processes.

5. 'It is amazing how many metal ions are present in the cells and fluids of the body.' Discuss this statement and give examples of places where these metals ions are used.

6. What are the main differences between the fluids inside and outside the cells of our body? How do their contents determine the functions of the body? Give one example of the importance of the movement of ions through the membranes of the cells.

7. What evidence is there that root crops that pick up elements and ions from the soil are essential for a balanced diet and healthy living? What would happen if we did not take in sufficient quantities of micro quantities of these essential elements? Discuss examples of three such ions.

8. In the light of the information about the action of zinc ions and also bromide ions on body processes, draw a cartoon to illustrate a possible use of zinc bromide (of course it might be poisonous!).

References

1. D. Derbyshire. Cadmium linked to breast cancer. *Daily Telegraph*, 14 July 2003, 6.
2. G. Rayner-Canham and S. Avery. Thallium – a poisoner's favourite. *Education in Chemistry*, September 2003, **40**(5), 132.
3. D. Bradley. Chemical channels win 2003 Nobel prize. *Education in Chemistry*, January 2004, **41**(1), 6.
4. G. Cragg and D. Newman. Nature's bounty. *Chemistry in Britain*, January 2001, 22ff.
5. P. McGowan. Cancer chemotherapy gets heavy. *Education in Chemistry*, September 2001, **38**(5), 134.
6. 'Perspectives'. Attacking cancer with a light sabre. *Chemistry in Britain*, July 1999, 17.

8 Water

Learning objectives

- To draw together various items of chemistry that reveal the uniqueness of water for metabolic processes.

- To consider the properties of water vapour and steam.

- To briefly consider the uses of ice.

Diagnostic test

Try this short test. If you score more than 80 % you can use the chapter as a revision of your knowledge. If you score less than 80 % you probably need to work through the text and test yourself again at the end using the same test. If you still score less than 80 % then come back to the chapter after a few days and read again.

1. What is the chemical formula for water? (1)

2. Complete and balance this equation: $2H_2 + O_2 \rightarrow$ (1)

3. Complete this word equation: solute + solvent $\rightarrow$ (1)

4. Water forms two ions. What are they? (1)

Chemistry: An Introduction for Medical and Health Sciences, A. Jones
© 2005 John Wiley & Sons, Ltd

5. The process that allows only a solvent like water to pass through a semi-permeable membrane is called what? (1)

6. The process that allows water and small molecules and ions to pass through a membrane is called what? (1)

7. Soaps and detergents have something special about their molecules. What is it? (2)

8. Water is a small molecule and would be expected to boil at a low temperature. What makes it boil at the high temperature of 100 °C? (2)

Total 10 (80% = 8)
Answers at the end of the chapter.

'My battle with the bottle' was the headline in a newspaper.[1] The page contained a long article relating the problems that actor Anthony Andrews had from drinking too much water to keep his vocal cords lubricated. It almost killed him. Water is probably the most important chemical in the universe and equally important in our body and its thousands of metabolic processes.

The solubility of drugs and medicines in either water or fats plays an important part in administering the active components to the sites where they are needed. Water solubility is very important when drugs are taken orally as they usually have to dissolve in gastric fluids before being taken up and transported to the parts of the body where they are needed, for example soluble aspirin. Some drugs, when they reach the site, dissolve in the fluids of the cells and are used by the cells for treatment of the ailment.

The pH of the organ or cell environments also affect solubility and this can be used to dispense drugs to where they are most needed. Some drugs are

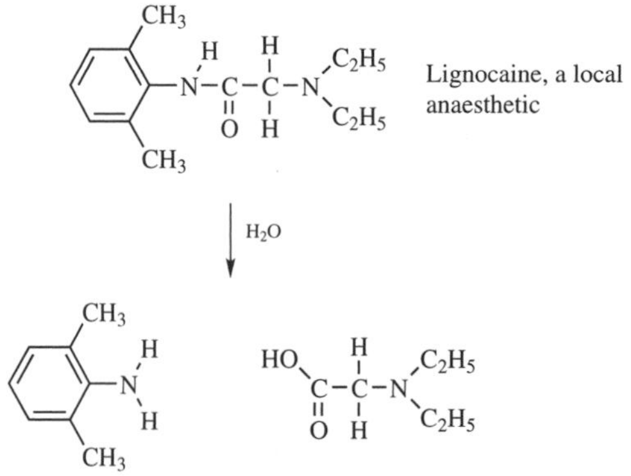

Figure 8.1 Hydrolysis of lignocaine

purposefully designed to be slowly broken up by water by hydrolysis. Lignocaine, a local anaesthetic used in dentistry, is slowly decomposed and metabolized by water (Figure 8.1). This allows the effect of the local anaesthetic to slowly wear off, but quickly enough not to have any lasting effects.

8.1 Introduction. What makes water so unique?

We drink water in many beverages, bathe in it, wash and clean ourselves in it, eat it as ice and breathe it in as its vapour. It forms part of the air, most of the sea and a considerable amount of the polar regions. It covers approximately three-quarters of the Earth's surface. It falls as rain, mist, fog, snow, hail, frost and ice. Our bodies are over 90 % composed of it and all the chemical reactions of the body depend upon it. Without it we would die – too much of it and we will drown.

Our bodies make it when they produce energy from oxidation of carbohydrates ready for us to use and live. The water vapour produced by this reaction is exhaled in breath. We make hundreds of litres of it in a year.

$$C_6H_{12}O_6 + 6O_2 \rightarrow 6CO_2 + 6H_2O \qquad (8.1)$$

Before we were born, we lived in a sac of it, without drowning, but we need special apparatus to breathe in it after we are born. Fish manage to extract enough air from it to breathe.

Deserts do not have enough of it but monsoons have too much of it. British holiday makers hate it but Sahara dwellers long for it. Skiers love it, but motorway drivers hate it.

It has a unique property of freezing at $0\,°C$ and boiling at $100\,°C$. It is a very important solvent but not everything dissolves in it. Fortunately our skin does not dissolve in it.

It makes us appreciate the beauty in a wide range of the scenery such as waterfalls, snow scenes and cloud formations.

You can win an Olympic medal by swimming in it, sailing on it, skating on it and sliding down hills on it. Water parks at holiday resorts are all the rage and full of excitement.

Water, ice, steam, water vapour – we are dependent upon it.

8.1.1   Water is special; is it unique?

So what is water and what makes it so special?

$$O_2 + 2H_2 \rightarrow 2H_2O + heat \qquad (8.2)$$

We can make it in the laboratory as it is a compound formed when oxygen gas reacts with hydrogen gas. The gas mixture needs either a spark or a flame to start the reaction. The chemical reaction, once started, is exothermic (gives out heat). The resulting water is a vapour but quickly condenses to a clear liquid at room temperatures.

In our body, water is made by our cells as a by-product of the oxidation by available oxygen of any carbohydrate fuel. Heat is given out in the reaction and is used to keep us warm.

Water is a unique compound of hydrogen and oxygen held together by covalent bonds. Between the molecules there is also an attraction of the hydrogen and the oxygen atoms. This is caused by residual small opposite charges on the H and O

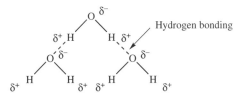

Figure 8.2 Hydrogen bonding

atoms. The attraction between these two slightly charged atoms causes a weak bonding between these atoms. It is usually shown as a dotted line between the atoms. This is called 'hydrogen bonding' (Figure 8.2). It is sufficiently strong to join up three or four water molecules together to form liquid water. These hydrogen bonds require energy to break them. H_2O on its own is a light molecule and would be a gas just like steam, but because of hydrogen bonding three molecules join up and cause the water to be a heavier molecule and become a liquid at room temperatures.

The residual small charge (δ^+ and δ^-) are caused by the fact that oxygen is on the right-hand side of the periodic table as compared with hydrogen. This means that the O atoms draw the electrons in the O—H bond slightly towards themselves and away from the H. So, relative to each other, O is slightly more negative than H, which is slightly positive. This means that the oxygens and the hydrogens in adjacent molecules, attract each other, forming hydrogen bonds. These bonds are also important for holding together parts of larger molecules in their shapes, including holding DNA into its spiral shape. In blocks of ice these hydrogen bonds are even more extensive and form a matrix throughout the ice block. To make steam from water the hydrogen bonds have to be broken by applying heat and forming single H_2O molecules that are light enough to escape into the air.

8.2 Chemical reactions in aqueous solution

When considering chemical reactions, we often take it for granted that the chemical reaction must take place in water, but there are other solvents for materials, including oils and fats. Water and oil do not mix and often something that is soluble in one is not soluble in the other. This is also true in human metabolism, where there are aqueous solutions and also oil/fat-based material, e.g. some vitamins (B and C) are water-soluble and others are fat-soluble.

The majority of chemical reactions inside the laboratory and also in cells occur in aqueous solutions. There are ions in aqueous solutions which are crucial for chemical reactions in cells. The sodium and potassium ions in solution are vital for the passage of messages from one cell to another. The building of cell proteins from small units of amino acids occurs in aqueous solution. Therefore it is essential that we regularly take in enough water either in liquid form or locked up in foods.

8.3 Dissolving and solubility: water is a great solvent

When a solute dissolves in a solvent then a true solution is made. There is a complete mixing of solute in a solvent:

$$Solute + solvent = solution$$

Each solvent is either covalent or ionic in nature and so are solutes. Generally ionic solvents will dissolve materials of similar nature, i.e. ionic solids. Similarly, covalent solvents will dissolve covalent organic compounds. Some groups in organic compounds, are water-loving, 'hydrophilic groups' and help the compounds to be soluble. These are groups like OH, COOH and SO_3Na groups. When wanting to synthesize a water-soluble drug, these groups are usually introduced.

Water is a slightly ionic compound and dissolves metallic salts, e.g. NaCl, and ionic compounds. Generally water will not dissolve organic covalent compounds unless they have a hydrophilic group present. This is just as well as we (as a collection of organic molecules) might dissolve in the rain!

Cell materials are either water-soluble or water-insoluble. The smaller building blocks of cells like amino acids and glucose are water-soluble, but the large molecules they eventually synthesize, like proteins, cell walls, fats and long-chain carbohydrates are water-insoluble. This is a subtle but significant fact. The water-soluble units can be carried around the body in aqueous solutions to sites where they are needed, but once a new larger molecules are synthesized then these are water-insoluble and so cannot be removed from the cell unless an enzyme or chemical attack takes place to break them up.

The possible ionic nature of the small molecules of amino acids means that they can form zwitter ions. The large number of OHs in glucose helps it to be water soluble, whereas the compounds it synthesizes, proteins and long-chain carbohydrates, are large molecules with molecular masses of greater than 20 000, and this prevents them from dissolving in water.

Although they do not dissolve, under the right conditions these compounds are chemically attacked by water (in weak acids in the presence of a catalyst called an enzyme). They are then broken down into the smaller original units. This reaction is called 'hydrolysis'. This chemical attack is not the same as the simple reversible physical change of dissolving. When something dissolves in water and a solution is formed, on evaporation of the water the original materials are left unchanged. This is not so in a chemical reaction – here the materials are changed into new compounds and the reaction is not reversible.

8.3.1 Solubility and dissolving

The solubility of any material in a given amount of water is a characteristic physical property of the material and is dependent upon the solvent used and the temperature. The higher the temperature the more a solid dissolves in the solvent. At any temperature the solvent will reach saturation point and will hold no more of the solid. On cooling, the excess solid comes back out of solution, usually in crystalline form, leaving the solution still saturated at that particular temperature.

Some of the most soluble materials in water are salt, sodium chloride and the smaller molecules of some sugars, including glucose and sucrose.

The degree of solubility in water can be expressed in a few different ways. The following are for solids dissolving in a solvent:

- percentage solubility – this is the number of grams of solute dissolved in 100 g solvent, sometimes expressed as grams of solute in a solvent (usually water) made up to 100 dm^3 (or grams) of solution;

- molarity – the number of moles of a solute dissolved in 1000 dm^3 of water; the concentrations of most solutions are expressed in this way as mol/dm^3 or mol/l (a more complete treatment of the quantitative aspects of molarity and concentration calculations is given in Chapter 15).

Because solubility depends upon the temperature of the solvent, it is usually referenced to room temperature, or 25 °C.

The solubility of gases in water is dependant upon temperature and atmospheric pressure and, providing the gas does not react with the water, follows Henry's law, which states that its solubility in the liquid is directly proportional to the partial pressure of the gas.

8.3.2 Too much water

The actor Anthony Andrews had no idea that drinking too much water could be harmful. Under the hot stage lights it is easy to become thirsty and even dehydrated, so between scenes he would go to his room and drink water, sometimes as much as 6 litres a day (his normal amount was about 3 litres). He was a very fit person but was putting himself under risk by over-drinking water and so suffering 'hyponatraemia' (this condition is more common with young people on ecstasy or even marathon runners, who take on drinks as they run). What happens is that the salts in solution in the body become more and more diluted, which results in nausea, headaches, weakness, confusion, unsteadiness, agitation, delirium and unconsciousness. These symptoms also similar to those of dehydration. The experience has made the actor a strong advocate of warning people not to become addicted to swigging mineral water, as water intoxication is a real possibility. Runners and exercise addicts should be careful of not over-grabbing water between tasks. It nearly killed Andrew. He returned to his role in 'My Fair Lady' a much more circumspect drinker . . . of water.

It is just as well that not everything is soluble in water, otherwise wood, metals, plastics, leaves, stones and skin would not have the advantageous properties they do in giving strength in structures and protection from weather, etc.

8.4   Osmosis

Osmosis is a process that allows the small solvent molecules, like water, to pass through a semi-permeable membrane but prevents the larger solute, such as sugar molecules, from passing through. Solvent passes from a more dilute solution (rich in solvent) through the semi-permeable membrane into the more concentrated solution (rich in solute).

Osmosis is an important process in any biological system, i.e. plants and animals, including ourselves. Many of the cell processes depend upon the ability of cell walls to act as semi-permeable membranes and allow the passage of fluids depending upon the concentrations of solutions inside and outside the cells.

To stop the osmosis occurring, the pressure P, in Figure 8.3, can be applied to the left-hand side. This pressure will be equal to the osmotic pressure exerted by the solution in the opposite direction. If the external applied pressure, P, is greater than the osmotic pressure then 'reverse osmosis' occurs and molecules can be forced to pass from the stronger to the weaker solution. In this process, the semi-permeable membrane acts as a 'molecular filter' to remove the solute particles. In some areas of the world this process is used for desalination of sea water, i.e. getting rid of salts from water. It is also used in emergency life raft survival kits to enable drinking water to be made from sea water.

The numerical value of the osmotic pressure depends upon the total concentration of the solute and thus the numbers of particles or ions present. When two solutions

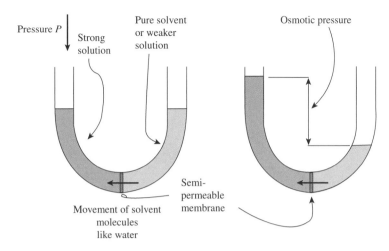

Figure 8.3   Osmosis

have the same osmotic pressure they are said to be 'isotonic'. Any solutions administered to the body intravenously must be isotonic with the body fluids. Notice how sports drink makers push this point in their adverts. It sounds 'technical' to say a drink is isotonic and this could help to influence the purchaser.

If cells (including blood cells) are immersed in solutions with a higher concentration of materials, then the osmotic pressures causes the water to pass from the cells and they shrivel. This is because water passes out of the cells through the cell walls, which are semi-permeable, into the more concentrated solution. This is called 'crenation' and can be disastrous for the cells. Food preservation processes can use this to advantage, for example, if meat is treated with salt then any bacteria cells on the surface shrivel and die. Similarly fruit can be covered with sugar with the same effect, and candied fruit is formed.

People who eat a lot of very salty food experience water retention in tissue cells, because water taken in as drinks to try to compensate for this enters the cells, which have more concentrated salt solutions, resulting in the appearance of puffiness which is called 'oedema'. It is worth looking up the remedy for this condition and possible medication.

8.5 Dialysis

The process of dialysis (Figure 8.4) in cells differs slightly from osmosis. In dialysis through a cell membrane, water and small molecules and ions pass through

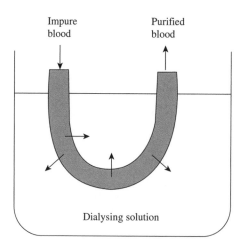

Figure 8.4   Dialysis; waste products dialyse out into the 'washing' dialysing solution

membranes. The larger proteins and cells cannot pass through. This important process is applied to artificial kidney machines, used to purify blood. Blood passes through a membrane of a cellophane tube and the tube is surrounded by a dialysing solution (formulated to contain all the ions of the blood and plasma). This solution contains all the ions in the same concentration as blood but none of the impurities. The small molecules of the undesirable impurities, such as urea, pass out of the blood, through the membrane, into the outer solution, and hence are removed. Without this artificial purification process some patients would die.

Haemodialysis usually takes about 4–7 h and the dialysing solutions are changed periodically. Ask in your hospital about the exact procedures used for dialysis, usually organized by the renal unit. You also might ask what stringent safety and health risk precautions the workers must undertake when working in the unit.

8.6   Colloids

You already know that soluble materials completely dissolve in a solvent to form a clear 'true solution', and the particles of the solute do not separate out from the solution. Also there are, in some situations, materials that form a 'suspension' of fine particles of a solid in a liquid. These are not true solutions and given time the particles slowly separate out, for example, when earth is stirred with water the fine earth particles suspended in water eventually fall to the bottom of the container. However, there is a group of materials called 'colloids' that is in between these two types of systems and the particle sizes are in between a true solution and a suspension. Whereas insoluble materials can be filtered, colloids cannot. However, they can be separated by dialysis.

Another way to separate out these fine particles is to use a high-speed centrifuge. This applies an artificial 'gravity' to the particles and cause them to be pushed to the bottom of the tube. In some circumstances the colloid is so fine that it must be 'coagulated' into larger particles before they can be separated, often by warming or adding an electrolyte or ionic solution. These particles are then large enough to filter off or they may fall to the bottom of a container or can be centrifuged. One example of a colloid is blood – you probably already know it has to be centrifuged to separate the red cells from the plasma.

8.7 Water, washing and detergents

Water is not miscible with all materials and it will not dissolve all materials. Greases, oils and fats are not water-soluble. To clean materials contaminated by these, soaps and detergents are used. These materials are chemicals that can dissolve oils and greases and are also water-soluble. Often these cleansing agents are carbon-based materials with long carbon chains (which are grease-soluble) and a water-soluble group at the end of its molecule. As a result they are able to mix easily with greases and oils and their water-soluble ending makes it possible for them to take the dirt and oils off the surface of a material and take it into a colloidal state in water due to its water-soluble or 'water-loving' head, and the impurities are washed away. These large grease–detergent colloidal particles are sometimes called 'micelles' (Figure 8.5).

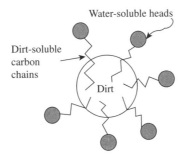

Figure 8.5   Structure of a micelle

Detergents can also be very effective in helping grease and fats to be made soluble in water. An example of a soap is sodium stearate. It has a COO^-Na^+ water-soluble end on a long carbon chain which is grease-soluble. Soaps can have a formula something like:

$$CH_3(CH_2)_{16}COO^-Na^+$$

Similarly, an example of a detergent is $CH_3(CH_2)_{20}C_6H_5SO_3^-Na^+$; here the $SO_3^-Na^+$ group is the water-soluble group. Notice that sodium ions are present. This also aids solubility as most sodium salts are water-soluble (see also Section 4.11).

Detergents also have the ability to lower the surface tension of water and 'wet' surfaces of materials to make them accessible to water and washing. Water behaves as though it has a thin skin over its surface that has to be broken to get below its surface (flies and beetles use this when they walk over the surface of water). The action of soaps and detergents breaks down this surface layer. A mixture of water plus soap or a detergent can wash a material more effectively than water alone.

8.8   Water vapour

We have mainly concentrated upon liquid water but the solid forms – ice, snow and hail – along with water vapour are also important items and worthy of further study. First we will briefly look at water vapour.

There is water vapour in the air, as can be seen on a cold day when it condenses out of the air of a room onto the window panes of a house (unless they are double or triple glazed), or are even seen as we breathe out. Water vapour comes from any process of burning a hydrocarbon in oxygen, and that includes us as we burn up glucose in our bodies to produce energy plus the waste gases of CO_2 and H_2O. Our metabolic processes also produce water and water vapour, which we exhale in our breath. Plants do a similar thing. Cars, when they burn petrol, and power stations as they burn oil, gas or coal all produce water vapour.

Water vapour is colourless, odourless and tasteless. Water vapour is not quite the same as steam. Water vapour is all around us now at room temperatures in about 0.5 % of the volume of air. The warmer the air, the more water vapour it can hold. In hot climates it is possible to get very high humidity as more water vapour is held in the air.

Steam is produced when water boils and water molecules are forced into the air. These molecules are very hot and in the area just above the liquid they appear colourless, but a little further away from the liquid they begin to cool off slightly and to condense into tiny droplets of water to form a mini-cloud which we call steam. Steam is actually hot droplets of water suspended in air. On a cold surface these easily condense to a liquid, leaving some of the normal water vapour in the air.

It is always amazing to see how much water vapour we have given out during the night through breathing. In my childhood, before central heating was common in houses, on cold winter mornings you could see your breath as you lay in bed plucking up courage to get washed and dressed. The window panes had wonderful ice patterns on the inside due to the ice crystals on the glass. We survived – just!

In centrally heated houses and hospitals sometimes an open dish of water is used in a room to evaporate and give more water vapour in the air. Air extraction and circulation systems or air-conditioning plants condense out the water vapour in their purification processes and so the air 'feels dry'. Flowers have a similar effect of adding some water vapour to the air by evaporation from their leaves and flowers.

Humid days are those when there is too much water vapour in the air and we feel uncomfortable. A similar effect occurs inside swimming baths or in the extreme case of sauna baths. This often makes people feel lethargic and lacking in energy. This can be a problem in some equatorial regions and offices are often air-conditioned. In air-conditioning plants the air is drawn through rooms filled with refrigerated pipes and blankets of air-purifying chemicals. These must be changed regularly as bacteria can breed in these moist places, particularly those bacteria that cause bronchial infections such as Legionnaire's disease, a form of pneumonia. Unless the system is cleaned regularly, these bacteria are blown all around the building with the 'freshened' air. This purification of circulated air is particularly vital in hospitals.

8.9 Evaporation from skin

Why is it that when you want to be cooled down, you wipe your forehead with water or a moist cloth? Why, if you wipe your arm or forehead with alcohol, ethanol, or better still with ether, it feels even cooler? For anything to evaporate it must have energy. Water, alcohol or ether on your hand takes the heat from your skin in order to evaporate, hence leaving your skin cooler. You can test this by putting a little piece of cotton wool over the bulb of a thermometer and dropping onto it a small amount of one of the liquids mentioned, then allowing it to evaporate or blowing on it to help evaporation. Watch the temperature on the thermometer. The heat needed for evaporation is called the 'latent heat of evaporation', and each material requires a different amount. The ether feels cooler as its boiling point and evaporation temperature are lower than that of water, so it evaporates rapidly and takes the energy from the skin rapidly. The boiling point of diethyl ether, $(C_2H_5)O$, is 34 °C, that of ethanol, C_2H_5OH, 78 °C and that of water, H_2O, 100 °C.

8.10 Solid water

I think I will have to go to the doctor as I have always got cold feet; is it a sign of poor circulation?

The various forms of solid water are ice, snow, frost and hail. Ice is used for cooling things down, in drinks and ice packs. Because, for ice to melt, it must take heat from somewhere. When we drink a cold drink the heat is taken from the mouth or throat and this lowers our temperature slightly.

Our bodies act in the same way as thermostats. They do not allow the temperature to go too low before some extra system switches on to compensate and restore the temperature, by shivering or burning up some carbohydrate fuel to give out heat. If you put an ice pack on a bruise, it brings the blood to the area, which compensates for the lowering temperature and in so doing helps the recovery rate as the blood does its work.

8.11 Hydrolysis

Hydrolysis is a chemical reaction in which water reacts with a molecule and decomposes it. There are many such reactions involving hydrolysis in the area of chemistry, but the most significant biological reaction is the hydrolysis of a protein.

The attack on a protein by water usually occurs in the presence of an acid and is catalysed by enzymes. The peptide bond breaks and the H adds on to one side of the bond and the OH of the water on to the other side of the broken bond (Figure 8.6).

$$HOOC\,CH_2 \cdot \mathbf{NH\,CO} \cdot CH_2 \cdot NH_2 + H_2O \rightarrow$$
$$HOOC \cdot CH_2 \cdot NH_2 + HOOC \cdot CH_2 \cdot NH_2 \tag{8.3}$$

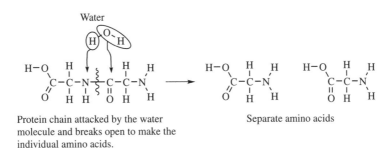

Water

Protein chain attacked by the water molecule and breaks open to make the individual amino acids.

Separate amino acids

Figure 8.6 Hydrolysis and breaking of a peptide link

So there we are ... water is so important and has many chemically and biomedically important roles.

Answers to the diagnostic test	
1. H_2O	(1)
2. $2H_2 + O_2 \rightarrow 2H_2O$	(1)
3. Solute + solvent $\rightarrow$ solution	(1)
4. H^+ hydrogen ions and OH^- hydroxide ions	(1)
5. Osmosis	(1)
6. Dialysis	(1)

7. One end of the molecule is water-soluble; the other end is soluble in
   grease   (2)

8. Hydrogen bonding holding three or four water molecules together   (2)

Further questions

1. Explain the vital role hydrogen bonding has in making H_2O so useful to living
   organisms.

2. Explain the difference between 'osmosis' and 'dialysis'.

3. Give some examples of materials that dissolve in the water of our bodies.

4. Describe a vital role for some electrolytes or ions.

5. Explain what is meant by a 'colloid' and give some examples. How can a colloid be
   separated from the liquid it is in?

6. Draw a table with three headings, Solid, Liquid and Vapour, and list the vital properties
   of the three forms of water.

7. Explain exactly what you mean when you say a solution is 'isotonic'. Why is this so
   important?

8. A very good 'ice pack' can be made if ice is mixed with salt inside a plastic bag. It can
   drop the temperature to $-10\,°C$.

   i. Explain the reason for this.
   ii. Why is salt added to icy roads in winter? Does the road temperature go up, stay the
       same or go down?

References

1. 'My battle with the bottle'. *Daily Telegraph*, 21 August 2003, 21.

9 Acids and Bases

Learning objectives

- To understand what is meant by the terms acid, base, alkali and pH.

- To appreciate the mechanism of buffer solutions.

- To see the relevance of acids and bases to body chemistry, including digestion.

- To introduce and revise the importance of amino acids and zwitterions.

Diagnostic test

Try this short test. If you score more than 80 % you can use the chapter as a revision of your knowledge. If you score less than 80 % you probably need to work through the text and test yourself again at the end using the same test. If you still score less than 80 % then come back to the chapter after a few days and read again.

1. What is the essential constituent of any acid? (1)

2. What is the general name of the compound produced when an acid reacts with an alkali? (1)

3. Complete this word equation: acid + base = (2)

4. Which of these are strong acids? Hydrochloric, sulfuric, carbonic, acetic? (2)

Chemistry: An Introduction for Medical and Health Sciences, A. Jones
© 2005 John Wiley & Sons, Ltd

5. What is a zwitter ion?   (1)

6. What is a buffer solution?   (2)

7. What pH range do acids have?   (1)

Total 10 (80 % = 8)
Answers at the end of the chapter.

'I feel ghastly, my tongue is like leather and my stomach is upset. It must have been that big meal I had last night and the few drinks I had. My stomach is out of order.'

'Then take an Alka-Seltzer or something. That helps. It is good for upset stomachs. It helps to restore the correct acidity.'

Buffers ... we all need them. These are chemicals that manage to maintain a stable body pH by the way they react with excess acid or alkali. Too much acid and a stomach ulcer develops; too little acid and digestion of foods is affected. Buffers keep the pH just right. Over-eating makes the stomach buffers work overtime and they can take some time to catch up after over-indulgence. The proteins and amino acids of our bodies work as part of the buffer system.

9.1 Acids

The term 'acid' has entered modern speech in terms such as 'acid rain', 'acid parties', 'acid indigestion', etc. These terms are used in conversational language but not in a scientifically correct way. In science we need to define clearly what is meant by 'acid'. Similarly, confusion of terminology has arisen with the word 'base', e.g. meaning low or bottom, as in the word 'basement'.

'Salt' is often used to mean common salt or sodium chloride, whereas in fact 'salt' is a general name used for a wide range of compounds produced when an acid reacts with an alkali. The developing definition of what is meant by 'acid' is traced in an article by John W. Nicholson.[1]

An acid is a substance containing hydrogen atoms, some of which, when the acid is dissolved in water, produce hydrogen ions. All acids have some properties in common:

- pH values lower than the neutral value of 7; $pH = -\log_{10} [H^+]$;

- a sharp taste (but you should never taste an unknown material to see if it is an acid because chemical tests are much better);

- strong acids can damage the skin and be dangerous, e.g. sulfuric acid;

- most acids are soluble in water and so release H^+ ions in solution;

- they can be neutralized by an equivalent amount of a base or alkali;

- they react with carbonates and bicarbonates to give off carbon dioxide gas;

- they react with some metals to release hydrogen gas;

- they are usually compounds of the nonmetallic elements C, N, S, P, O, Cl, e.g. HCl, H_2SO_4 or H_3PO_4.

9.1.1 pH and the log scale

For most solutions the concentration is expressed as the number of grams of the substance in a litre or 1 dm^3 of solution. Sometimes it is expressed as a molarity (M). Uniquely, acids and alkalis have their own notation, called pH. It is a scale that avoids the use of small numbers and powers of 10 and simply expresses the acidity of a weak solution on a scale of 1–14. To do this it converts the concentration of the solution into a number using the expression $pH = -\log_{10} [H^+]$.

The pH of a solution of say 0.001 M hydrochloric acid ($H^+ Cl^-$), with a hydrogen ion, H^+, concentration of 1×10^{-3} g/dm^3 is

$$pH = -\log_{10}[1 \times 10^{-3}]$$
$$= -[\log_{10}1 + \log_{10}10^{-3}]$$
$$= -[\log_{10}1 + (-3)\log_{10}10]$$

Remembering or looking up in log tables that $\log_{10}1 = 0$ and $\log_{10}10 = 1$, and removing the brackets,

$$= -0 + (-)(-)3$$
$$= 0 + 3$$
$$= 3$$

It is easier to say pH = 3 than to do the maths every time.

Values of pH from 1 to 6 are termed acidic and 7 as neutral, whereas alkaline solutions have values of 8–14. Coloured solutions called 'indicators' have been specially made to change colour when added to solutions of different pH values. This colour change is a rough indication of the pH of a solution, and paper strips containing these dyes can be used as test strips to check the pH of solutions.

9.1.2 Do not confuse strong acid with concentrated or weak acid with dilute

There are various ways to classify acids. One is to consider the number of ionizable hydrogen atoms in a molecule. For example one H in a molecule that can form hydrogen ions is called a *mono basic acid*, e.g. $H^+ Cl^-$. A *dibasic acid* will give two H^+s, e.g. H_2SO_4. An example of a *tribasic acid* is phosphoric acid, H_3PO_4.

A further way to characterize an acid is to say if it is a strong or weak acid, but a 'strong acid' does not mean a 'concentrated acid'. Concentrated means a large quantity of the acid dissolved in a solution. Dilute means a small quantity of a substance dissolved in a solution. A *strong* acid, when dissolved in water, contains molecules that almost *entirely* dissociate (or ionize), producing a huge number of hydrogen ions. These hydrogen ions in solution give a low pH (usually 1 or 2). It is these hydrogen ions that give acids their characteristic properties, e.g. hydrochloric acid, HCl, and sulfuric acid, H_2SO_4. HA is the formula of a typical acid:

$$HA \rightleftharpoons H^+ + A^- \tag{9.1}$$

$$HCl \rightleftharpoons H^+ + Cl^-$$

$$H_2SO_4 \rightleftharpoons 2H^+ + SO_4^{2-} \tag{9.2}$$

A *weak* acid also contains hydrogen atoms in the molecule of the acid but produces few hydrogen ions and so has a relatively high pH, nearer the neutral value of 7 (usually 5 or 6). The weak acids are reluctant to dissociate, or ionize, in solution. There are more HA undissociated molecules than hydrogen ions. Examples include the 'organic' molecules of ethanoic acid (acetic acid, CH_3COOH), citric acid, carbonic acid (H_2CO_3), lactic acid and amino acids. These only give small quantities of hydrogen ions in solution as compared with the number of undissociated molecules. It is the number of hydrogen ions present in solution that determines the pH. The predominant direction of the equilibrium is towards the undissociated acid molecule:

$$HA \rightleftharpoons H^+ + A^-$$
$$CH_3COOH \rightleftharpoons CH_3COO^- + H^+ \tag{9.3}$$

A concentrated solution of a weak acid still only produces a small number of hydrogen ions (higher pH). Similarly, a dilute solution of a strong acid is still able to almost completely ionize and give a large number of hydrogen ions (lower pH). Therefore it is possible to get a concentrated solution of a 'weak' acid or a dilute solution of a 'strong' acid, but the strong acids produce more hydrogen ions in solution than the weak acids.

9.1.3 Hydrogen atoms in molecules of acids

Not all the hydrogen atoms contained in the molecules of organic acids ionize when in solution to give H^+ ions. It is only the 'acid active' hydrogens that can do this.

This particularly applies to the weaker organic acids listed above. For example, in ethanoic acid, CH_3COOH it is only the **H** of the COOH that is ionizable; all the Hs on the CH_3 remain attached to the C at all times.

$$CH_3COOH \rightleftharpoons CH_3COO^- + \mathbf{H}^+ \tag{9.4}$$

Similarly for amino acids it is only the H of the COOH that ionizes to give a H^+ ion:

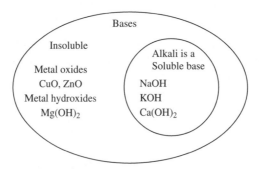

$$NH_2CH_2COOH \rightleftharpoons NH_2CH_2COO^- + \mathbf{H}^+ \tag{9.5}$$

Most of the acids we have in our bodies are produced as a result of the reactions going on inside the cells and only a small quantity is taken in as part of our food and drink. To make sure this externally added acidity does not grossly affect the mechanisms of the body cells and disturb the static state or 'homeostasis', the body has an in-built regulator or 'buffer' system.

9.2   Bases and alkali

The term 'base' is widely used in chemistry but in biology and medical science it means something more specific (Figure 9.1). In chemistry what we mean by 'base' is a material that is the opposite to an acid. They are compounds of a metal: either the oxide or hydroxide. Those bases that are soluble in water are usually termed 'alkalis' and include sodium hydroxide, NaOH, and potassium hydroxide, KOH.

Figure 9.1   Bases and alkalis

Uniquely, the solution of ammonia in water (the compound ammonium hydroxide, NH_4OH) is also an alkali. The insoluble oxides of copper (CuO) and iron (Fe_2O_3) are bases, but because they are insoluble in water they are not called alkalis.

The organic compounds called amino acids are unusual. They have an acid group at one end of the molecule (COOH) and a base group (NH_2) at the other. We call a molecule that has both acid and base properties 'amphoteric'.

9.2.1 Properties of bases in chemistry

- The pH of the solution lies between 8 and 14.

- Soluble bases are called alkalis, e.g. sodium hydroxide, NaOH, potassium hydroxide, KOH, and ammonium hydroxide, NH_4OH.

- Bases are oxides or hydroxides of metallic elements

- Bases and alkalis will react with acids to neutralize them, forming salts plus water:

$$Acid + base \rightarrow salt + water$$
$$Acid + alkali \rightarrow salt + water$$

- All alkalis contain a hydroxide ion, OH^-, that will react with and 'pick up' a H^+ ion to form a water molecule:

$$OH^- + H^+ \rightarrow H_2O \tag{9.6}$$

9.3 Bases containing nitrogen

These are best called 'nitrogenous bases', which is a term often used in relation to amines, amino acids and proteins. Bases referred to in medicine and biology usually contain nitrogen atoms that have the ability to pick up a proton and become a positive ion:

$$NH_3 + H^+ \rightarrow NH_4^+ \tag{9.7}$$
Ammonia

$$HOOCCH_2NH_2 + H^+ \rightarrow HOOCCH_2NH_3^+ \tag{9.8}$$

Amino acid here acts as a base as the nitrogen picks up a proton.

The bases contained in DNA are quite complex units which are built into the double inter-linked helix molecule of DNA. There are four main types of base, A, T, C and G, standing for adenine, thymine, cytosine and guanine, (see also Chapter 5), but they can also be protonated, i.e. pick up an H^+, by reacting with an acid.

9.4 Amino acids and zwitterions

Amino acids can occur as both acids and bases depending upon the nature of the solution or environment. This is because such compounds are unique in biochemistry. They are able to give H^+ ions (as all acids do) from the end of the molecule containing a COOH group, but also pick up H^+ at the other end of the molecule using the N atoms of the NH_2 group. If the correct acid/alkaline conditions are chosen, then these molecules can be both positively and negatively charged at the opposite ends of a single molecule. This is known as a zwitterion (zwei in German means two). (See also Chapter 5.)

$$^-OOCCH_2NH_2 + H^+ \rightarrow \mathbf{HOOCCH_2NH_2} \xrightarrow{+H^+} HOOCCH_2NH_3^+$$
$$- H^+ \text{ from COOH} \rightarrow {}^-OOCCH_2NH_3^+$$
$$\text{zwitterion}$$

$$^-OOCCH_2NH_2 \xleftarrow{-H^+} {}^-OOCCH_2NH_3^+ \rightarrow$$
$$\text{zwitterion}$$

$$\xrightarrow{+H^+} HOOCCH_2NH_3^+ \xrightarrow{-H^+} \mathbf{HOOCCH_2NH_2}$$

$$(9.9)$$

9.5 Salts

Salt is the general name of a set of compounds that are formed when an acid reacts with, or is neutralized by, a base or alkali. Common salt, sodium chloride, is only one of many thousands of salts.

In some cases it might be necessary in an equation to show if the substance is in aqueous solution or even in the solid state. To show these differences, subscripts or brackets are often used to help clarify the situation. Aqueous solutions are denoted by (aq) and the solid state by (s), l meaning liquid as in liquid water:

$$\text{Acid} + \text{base} \rightarrow \text{salt} + \text{water}$$
$$HCl(aq) + NaOH(aq) \rightarrow NaCl(aq) + H_2O(l)$$
$$H_2SO_4(aq) + 2KOH(aq) \rightarrow K_2SO_4(aq) + 2H_2O(l)$$
$$CH_3COOH(aq) + NaOH(aq) \rightarrow CH_3COONa(aq) + H_2O(l)$$
$$H_2SO_4(aq) + CuO(s) \rightarrow CuSO_4(aq) + H_2O(l)$$

$$(9.10)$$

In most equations it is assumed that the reaction is in aqueous solution and the (aq) etc. are omitted.

Salts can also be formed by the neutralization of the acid with other basic materials, including metal carbonates, bicarbonates and metal oxides.

Acid + metal carbonate (or bicarbonates) → salt + water + carbon dioxide gas

$$2HCl + Na_2CO_3 \rightarrow 2NaCl + H_2O + CO_2 \qquad (9.11)$$

You might read on packets of some 'anti-acid' stomach settlers and remedies that they contain aluminium or magnesium salts to help settle the stomach after over-eating and indigestion. The acid in the stomach has the main task of dissolving foods by the action of stomach acid, hydrochloric acid. If you over-indulge in food or drink, this system becomes over-worked. It desperately tries to make more and more HCl. In doing so it can over-produce HCl and cause acid indigestion. Many indigestion tablets contain either sodium bicarbonate or aluminium compounds. In Victorian times potions containing 'Bismuth' were used but these have been shown to be poisonous in too large a dose.

All sodium, potassium and ammonium salts dissolve in water. Look at the instructions and contents of a medicine or food and you will see what salts are present to make the material water soluble and so make it more easily absorbed in the stomach.

9.6 Neutralization

This is the process whereby the quantity of an acid is just balanced by the addition of a base or alkali. The hydrogen ions are just balanced by the OH^- ions in an aqueous solution to form neutral water

$$H^+ + OH^- \rightarrow H_2O \qquad (9.12)$$

9.7 Buffer solutions

Most chemical reactions occurring in our bodies work best in a specific pH range. Blood, for example, works at pH 7.4 and any variation of 0.2 units either way would render the person seriously ill. To make sure the pH values are kept at the appropriate best working values, our body cells employ a series of solutions called 'buffers'. These are molecules that resist any changes of acidity or alkalinity. Buffer solutions are a mixture of substances that interact with any incoming acid or alkali impurity to render them ineffective and help restore the pH of the solution to its original value.

You can design artificial mixtures of substances to make up buffers to keep solutions at any given pH by regulating its composition. Commercial 'stomach acidity regulators' bought in the chemist's shop have their own mixtures of chemicals designed to do this. The body has its own special system and is only interested in keeping the pH at the most effective working values. This is known as 'homeostasis'. The body's buffers, mainly proteins and amino acids, must have a wide flexibility to take care of any 'alien' impurities from causes ranging from disease to over-eating.

What are 'buffers' made of? First we will look at the body buffers of proteins designed to help maintain the correct metabolic conditions and pH.

9.7.1 Protein buffers

Proteins are the most abundant materials in the body. Proteins have long chains of carbon compounds, as many as 1000 or more, and have amino acid side chains sticking out of them. One protein of general formula could be $^+NH_3(CH_2)_n$ COOH:

$$^+NH_3(CH_2)_n COOH \xrightarrow{+H^+} NH_2(CH_2)_n COOH \xrightarrow{-H^+} NH_2(CH_2)_n COO^- \qquad (9.13)$$

An adverse external influx of acids could be removed by the protein buffer soaking up the H^+ ions to form a positive ion, hence restoring the original pH.

If OH^- ions are present, these react with the buffer, causing the amino acid or protein to make H^+ ions. These react with the hydroxide ions to form neutral water. More of the protein will ionize if more OH^- ions are added as an impurity. The above equilibrium moves to the right and the original pH is restored.

9.8 Buffers in the body

Haemoglobin (a compound of iron and proteins) buffers the blood system using the proteins present. This is essential for controlling the pH of the blood, which is necessary due to the uptake of acidic CO_2 gas formed when cells use carbohydrates, glucose, to give energy.

Carbonic acid is formed when carbon dioxide reacts with the water present in the cells.

$$CO_2 + H_2O \rightarrow H_2CO_3 \rightarrow H^+ + HCO_3^-.$$

This in turn produces H^+ ions, which have to be dealt with and neutralized by the buffer system to keep the pH at the best working value.

The oxygenated blood, which is slightly more alkaline, must transport the oxygen to the cells and the pH must not affect the workings of the other materials in the cell. The proteins buffers the solution to maintain the working pH of the cells, usually about pH 7.4, with a range of 7.35–7.45. It can be seen from this that any build up of CO_2 caused by breathing difficulties (e.g. emphysema) can grossly affect the sensitive buffering system and even overload it. If blood pH falls below 7.0, called 'acidosis', there is a severe depression of the nervous system and the person becomes disorientated and can go into a coma and die unless the pH is restored.

On the other hand, if the blood pH goes above 7.45 and blood CO_2 falls, causing hyperventilation, then the kidney attempts to compensate by decreasing the excretion of H^+ ions. A similar effect is caused by oxygen deficiency (e.g. high altitude sickness, brain damage or aspirin overdose). The simple remedy is to get the person to breathe into a paper bag, then rebreathe the exhaled air containing a larger proportion of CO_2 (but still enough oxygen), so increasing the acidy of the blood and lowering the pH.

The proteins in the blood (namely that of the haemoglobin and particularly the histidine and cysteine amino acids of the proteins) are excellent buffers and keep the blood working at its most effective pH of 7.4.

9.9 Digestion and acid attack

After eating too much food or drinking too much alcohol, the stomach sometimes becomes overloaded with the tasks it has to perform, particularly that of maintaining a steady constant working pH. The buffer system becomes overloaded and overworked and an excess of stomach acid builds up. This can lead to indigestion.

The re-stabilizing of the stomach pH can be helped by taking stomach powders: 'Rennies', 'Alkaseltzer' or small amounts of 'bicarbonate of soda', etc. These materials add a small amount of basic OH^- ions or bicarbonate, HCO_3^-, ions in order to react with the excess acid. Unfortunately the bicarbonate ions produce CO_2 gas which causes you to burp.

$$H^+ + HCO_3^- \rightarrow H_2O + CO_2 \qquad (9.14)$$

9.9.1   Other meanings of acid

The word 'acid' has been associated with the drug scene through 'acid parties'. The acid being referred to in this case is LSD, a hallucinatory drug which also has pain-relieving properties.

9.10   Acids in the environment

Carbon dioxide is a naturally occurring acid gas which on dissolving in water or moisture forms 'carbonic acid', the originator of the salts called carbonates. Carbon dioxide is generated in many places naturally, domestically and industrially. All the carbon in our bodies and plants comes indirectly from the carbon dioxide in the air (see Section 2.8). The greenhouse effect is partially caused by an increase of CO_2 in the air. The air allows ultraviolet light from the sun to pass through and reach the ground. Here it can be either absorbed or converted into a different energy form, such as heat. The CO_2 absorbs the energy, which prevents it escaping back into space. Therefore, heating of the Earth occurs as CO_2 concentrations in the air increase. The increases look marginal in terms of percentages, but are enough to influence our finely balanced living organisms and surface temperatures.

Sulfur dioxide is an acid gas and is an unfortunate by-product of the burning of fossil fuels, which often contain some sulfur. Filters and catalytic converters remove this from exhaust fumes from cars. Low-sulfur fuels are being marketed to help alleviate this problem. Power stations also try to filter sulfur out by mixing the waste gases with limestone (calcium carbonate). The calcium sulfate so formed is used as a soil conditioner in agriculture or sold to works making plaster boarding for the building industry.

Sulfur dioxide in the air dissolves in the rain, along with some of the oxygen present, forming a weak solution of sulfuric acid:

$$S + O_2 \rightarrow SO_2$$
$$2SO_2 + O_2 \rightarrow 2SO_3 \qquad\qquad (9.15)$$
$$SO_3 + H_2O \rightarrow H_2SO_4$$

The sulfuric acid attacks limestone walls and marble statues.

Burning fuels in air (oxygen and nitrogen) also makes small quantities of the oxides of nitrogen, NO_2. This dissolves in rain to form weak nitric acid. This also occurs naturally when lightning passes through the oxygen/nitrogen of the air and its solution in rain helps to restore the nitrates in the soil. Nitrates make excellent fertilizers.

Rain is naturally slightly acidic (due to CO_2 from plants and animals), but in recent years this has been added to by the excessive burning of fuels (cars, power stations, homes). Too high a concentration of all these acids in the air can kill plants and trees.

It must also be remembered that the eruption of volcanoes produces millions of tons of these acid gases each year, so acid pollution is not entirely our fault!

Much more can be said on this subject, but space does not allow us to do so. Books on environmental science cover this topic in more detail. Our planet is a finely balanced system that needs our help to maintain its equilibrium.

Answers to the diagnostic test

1. Hydrogen ions, H^+ (1)

2. Salt + water (1)

3. Acid + base = salt + water (2)

4. Hydrochloric, strong; sulfuric, strong; carbonic, weak; acetic, weak (2)

5. A molecule that possesses both an acid group and a basic group in a single molecule, e.g. an amino acid (1)

6. A solution that controls the pH of a solution (2)

7. Values below pH 7 (1)

Further questions

1. Put into equation form the following statements:

   i. Sulfuric acid, H_2SO_4, is a strong acid and when in aqueous solution it dissociates almost completely into hydrogen ions and sulfate ions.
   ii. Potassium hydroxide, KOH, is a strong base that is also soluble in water and so is called an alkali. It is almost completely dissociated in aqueous solution into its component ions.

iii. Sulfuric acid reacts with potassium hydroxide, in aqueous solution, to form a salt, which is itself fully ionized, potassium sulfate.

iv. Hydrochloric acid, HCl, reacts with solid calcium carbonate, $CaCO_3$, limestone, to form a salt and water. Carbon dioxide gas is given off.

v. Carbonic acid and ethanoic (CH_3COOH) acid are both weak acids and are predominantly in the molecular form, although carbonic acid is easily decomposed at room temperatures into carbon dioxide and water.

vi. Glycine is an amino acid, $NH_2 \cdot CH_2 \cdot COOH$. This molecule in acid solution forms a positive ion whereas in alkaline solution a negative ion is produced.

vii. Amino acids incorporated into proteins can act as pH controllers or 'buffers' to modify changes in acidity or alkalinity.

2. To keep our stomach at a constant working pH, buffer solutions can be used. Give one example. What materials do the body tissues and cells use to keep the pH constant.

3. What is meant by pH and how is it measured? What do all acids have in common?

4. 'Acids' are often thought of as being something bad. Comment upon this with arguments for and against.

5. Explain what is meant by 'base' when used in chemistry. How does a base differ from an alkali?

6. A volume of 20 cm^3 of sodium hydroxide solution is neutralized exactly by 40 cm^3 of 2 M hydrochloric acid. What is the molarity of the sodium hydroxide? 0.5, 1, 2, 4 or 8 M, you select.

7. If a DC electric current was applied to an acid solution of glycine, to which electrode would you expect the amino acid ions to migrate? Would it make any difference if the glycine was in acid, neutral or alkaline solution?

8. Rain is naturally slightly acidic but is made more acid by burning fuels and can contain carbonic, sulfuric and nitric acids. Explain how they are formed.

References

1. J. W. Nicholson. A brief history of acidity. *Education in Chemistry*, January 2004, **41**(1), 18–19.

10 Oxidation and Reduction

Learning objectives

- To draw together the ideas of redox reactions.

- To explain redox reactions in terms of electron shifts.

- To see the significance of redox reactions for metabolic processes.

Diagnostic test

Try this short test. If you score more than 80 % you can use the chapter as a revision of your knowledge. If you score less than 80 % you probably need to work through the text and test yourself again at the end using the same test. If you still score less than 80 % then come back to the chapter after a few days and read again.

1. Define oxidation with reference to oxygen (1)

2. Define reduction with reference to electron flow (1)

3. $CuO + H_2 \rightarrow Cu + H_2O$

   What has been oxidized to what? (2)

   What has been reduced to what? (2)

4. What is SOD? (1)

Chemistry: An Introduction for Medical and Health Sciences, A. Jones
© 2005 John Wiley & Sons, Ltd

5. What is the molecule NO called? (1)

6. Why are aspirin given after a heart attack? (2)

Total 10 (80 % = 8)
Answers at the end of the chapter.

'Students test viagra on Bolivian mountain', read the headline. The project, designed by Edinburgh University, took 110 students up 16 000 feet on Mount Chacaltaya, Bolivia, and gave them viagra for a week. Because the drug was known to relax blood vessels in the lungs, it was hoped that the drug would help to prevent high-altitude pulmonary oedema (fluid gathering on the lungs).

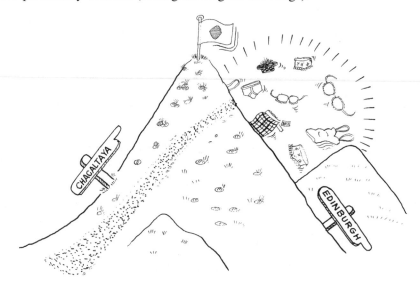

Viagra was one of the biggest selling drugs in the last few years; its action depends on the presence of minute quantities of an oxide present in our systems. Read on.

10.1 Definitions of oxidation and reduction

We use the terms reduction and oxidation in many different contexts, but the chemical ones need to be carefully defined.

1. Oxidation is the addition of oxygen to an element or compound or the removal of hydrogen from a compound. Reduction is the opposite of this, i.e. the addition of hydrogen or the removal of oxygen.

2. Oxidation is also applied to a process of electron loss and is always accompanied by reduction in another compound in the same reaction. Reduction is a process of electron gain.

3. The material that does the oxidizing is called the 'oxidizing agent'. When it oxidizes something, it is itself reduced. A reducing agent itself becomes oxidized when it reduces something.

We will be using many chemical equations in this chapter to show what is going on in a chemical reaction. You will not be expected to balance equations or remember the numbers in the equations. You will only need to understand the changes occurring and the principles of what is going on in the reactions.

10.1.1 Examples of the definitions

Definition number 1

When black copper oxide is heated and hydrogen gas is passed over it the copper oxide loses its oxygen to form copper metal and so is reduced, but notice that the hydrogen gains oxygen to become water, so it is oxidized:

$$CuO + H_2 \rightarrow Cu + H_2O \qquad (10.1a)$$

The oxidizing agent here is the copper oxide (as it supplies the oxygen) and so it is reduced to copper. The reducing agent here is hydrogen and so itself becomes oxidized to water.

When iron and steel are made by mixing coke (carbon) with the iron oxide ore and heating the whole lot in a blast furnace, then the following occurs:

$$Fe_2O_3 + 3C \rightarrow 2Fe + 3CO \qquad (10.1b)$$

The carbon is the reducing agent as it takes away the oxygen from the iron ore and itself is oxidized to carbon monoxide. The iron oxide has been reduced to iron.

Definition number 2

This second definition is useful when ion and electron movement is involved in chemical reactions. Chemical reactions involving ionic compounds are best interpreted by this definition. Consider the attack of dilute hydrochloric acid on metallic zinc. The zinc dissolves and forms zinc chloride solution and the hydrogen gas fizzes off:

$$Zn + 2HCl \rightarrow ZnCl_2 + H_2 \tag{10.2}$$

Because there is no oxygen involved in this reaction, it is difficult to see what is happening to the zinc. The second definition is more helpful in interpreting this reaction. Looking at the ions present:

$$Zn + 2H^+2Cl^- \rightarrow Zn^{2+} + 2Cl^- + H_2 \tag{10.3}$$

Zinc metal (Zn) loses electrons to form the positive ion, Zn^{2+}. The zinc is oxidized by the process of electron loss. The hydrogen ions, H^+, gain electrons to become neutral hydrogen gas (H_2), so the hydrogen ions are reduced. The chlorine ions do not change and are called 'spectator ions', so the equation could be simplified to show only the things that change:

$$Zn + 2H^+ \rightarrow Zn^{2+} + H_2 \tag{10.4}$$

Notice that the equations still balance with the same number of atoms and the same number of charges on both sides of the equation.

Iron is an element that can form compounds in two ionic states: Fe^{2+} and Fe^{3+}. These two states are involved in the transport of oxygen in the haemoglobin of the blood. The change of Fe^{2+} to Fe^{3+} means that the iron has lost one electron and the Fe^{2+} has been oxidized:

$$Fe^{2+} - e^- \rightarrow Fe^{3+}$$

Sometimes the two different oxidation states are indicated by the use of Roman numerals after the formula, e.g. $Fe(II)^{2+}$ and $Fe(III)^{3+}$. These Roman numerals are called 'oxidation number'. When the oxidation number of an element increases, it has been oxidized. A decrease in the oxidation number shows the element has been reduced. A further method of balancing redox equations is based on these oxidation numbers, but the concepts will not be pursued further in this book:

$$Fe(II)^{2+} - e^- \rightarrow Fe(III)^{3+}$$

Obviously some other chemical must supply the action of removing the electrons from Fe^{2+}. The reverse action is also possible and the $Fe(III)^{3+}$ can be reduced to $Fe(II)^{2+}$ in a different environment by the action of a reducing agent.

Redox is the general name given to the changes taking place when oxidation and reduction take place in chemical reactions. Because electrons are transferred from one atom or group to another, these reactions can be followed using electrical detection devices. Sometimes the tendency of materials to be oxidized or reduced can be expressed in terms of redox potentials and measured in volts.

10.2 Burning and oxidation

When anything burns in air or oxygen, it is by a process of oxidation/reduction. Natural gas CH_4 burns in air:

$$CH_4 + 2O_2 \rightarrow CO_2 + 2H_2O - \Delta H \qquad (10.5)$$

giving out energy to the surroundings. The methane has been oxidized to carbon dioxide and water. The oxidizing agent, oxygen, has been reduced to water.

The 'burning' of the body's fuel, glucose, involves oxidation/reduction reactions:

$$C_6H_{12}O_6 + 6O_2 \rightarrow 6CO_2 + 6H_2O - \Delta H \qquad (10.6)$$

giving out energy or heat to the body.

The term $\pm\Delta H$ at the end of some equations helps us to see what is going on in terms of energy changes in chemical reactions. The negative sign, $-\Delta H$, shows that heat has been lost from the reaction to the surroundings. This means that the chemical reaction is giving out heat and is said to be exothermic. A positive sign, $+\Delta H$, would mean that the reaction requires heat to occur and be maintained, and this is called an endothermic reaction.

10.3 Some applications of redox reactions to metabolic processes

10.3.1 Super oxides

In recent years newly invented detection devices have enabled scientists to investigate the minute quantities of materials that are involved in our metabolic processes. The 'super oxide' anion (O_2^-) is one of the enemies of the cells of our bodies. It is generated as a by-product of respiration with oxygen and is also made by specially activated white blood cells. In normal healthy people the body has its own way of disposing of these ions using of an enzyme called super oxide dismutase, or SOD. You can guess what this material says to the unwanted by-products!

After heart attacks and strokes or inflammatory diseases like arthritis, our bodies produce extra amounts of these super oxide ions. These can sometimes overwhelm the resources of SOD. This excess super oxide can then damage surrounding tissues. It is important therefore to reduce this concentration as quickly as possible.

The normal mode of action of the SOD is to catalyse the conversion of super oxide to hydrogen peroxide and molecular oxygen. The H_2O_2 (hydrogen peroxide) is decomposed into water and oxygen by another enzyme present in the blood, called peroxidase. If the super oxide is left too long in the system, it also reacts with the small quantities of nitric oxide present to produce the cytotoxic agent peroxynitrite, $[O—O—NO_2]^-$. This is particularly harmful to cells. Nitric oxide is also important to us as it is a key material used by the body to control blood flow and reduce the adhesion of platelet cells involved in blood clotting and thrombosis, so it must be preserved in the correct quantities.

> In 1998 my uncle was right in the middle of a vigorous game of squash. He collapsed with a heart attack. He would have died had it not been for Fred, his squash partner, who gave him two aspirins. He knew that aspirins help to soak up the excess super oxides produced by the heart attack. I always carry some aspirins now; well, at my age you never know.

Heart attacks can lead to excess super oxide production, which in turn saturates the SOD system; the excess super oxide then has a chance to react with nitric oxide and obstruct its useful role in keeping the blood flowing. This leads to a greater chance of blood clotting and a thrombosis. The most convenient remedy is to take something straight after the heart attack that thins the blood. This often takes the form of an artificial SOD medicine or injection. The more common remedy is to take two aspirins.

Recent research[1] in this area has been focused on mimicking the SOD chemicals and manufacturing drugs that replace the lost SOD and so reducing the super oxide formed after a heart attack.

10.4 Nitric oxide, NO or N(II)O

The electronic structure of nitric oxide, NO (Figure 10.1), is not as straightforward as some compounds we have discussed earlier in this book. It appears to have an 'odd' electron left over and does not obey the 'octet' rule for atoms in chemical bonding.

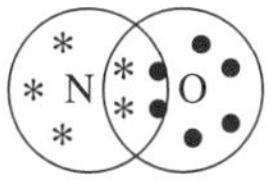

The oxygen has a
complete octet, but the
N only has seven electrons

Figure 10.1 Nitric oxide

Nitric oxide is an unstable gas and, if exposed to the air, it quickly oxidizes to
nitrogen dioxide, NO_2 (a brown gas). This helps it to overcome its 'odd' electron
unstable structure. However, in the closed environment of the body, this process
does not occur and so the NO carries out its unique roles within the body's cells. Its
production is probably a result of oxidation of amino acids or proteins. Nitric oxide
has been implicated in the biochemistry of virtually every mammalian organ system,
including inflammatory and degenerative diseases. It is present in the blood and
other parts of the body and there are very small quantities in the brain. These
concentrations are so small that it is only in recent years that it has been discovered
following the development of sensitive analytical instruments.

Other roles that have recently been discovered for nitric oxide include:

- The cholesterol-lowering statin drugs, e.g. fluvastatin, improve the blood flow,
  probably by increasing the amount of nitric oxide in arterial linings and so
  helping them to dilate.[2] (Do an internet search for nitric oxide to find recent
  research documents.)

- Nitric oxide has been found to be an important cell-signalling molecule and
  research is continuing into its role in causing migraines and brain damage after
  heart attacks (due to causes described above, linked with SOD).[2]

- The inhaling of small quantities of nitric oxide can help premature babies (or
  babies with holes in the heart) to breathe and prevent serious lung disorders.

- The 1998 Nobel Prize for medicine was won by Dr Robert Furchgott for his work
  on the roles of nitric oxide as a signalling molecule in the cardiovascular system.
  He said 'we know that nitric oxide acts as a signal molecule in the nervous
  system, as a weapon against infections, as a regulator of blood pressure and as a
  gate keeper of blood flow to different organs'.

- The discovery of its role as a useful chemical of the body has also led to the
  development of the anti-impotency drug viagra.

- Its involvement in a pathway for cancer therapies is being investigated.

Alfred Nobel (1833–1896) invented dynamite, which contains nitroglycerine. Nitroglycerin, in small quantities, was known even in his time to help alleviate chest pains caused by heart attacks. It is now known that it works by releasing nitric oxide into the body and particularly the coronary artery, resulting in dilation and a relief of chest pains by allowing more blood to pass. Alfred Nobel suffered from chest pains but refused to take nitroglycerine as he said it gave him a headache. He died of a heart attack.

10.5   Oxygen gas

Oxygen is essential for human life as each cell requires a constant supply to stay alive. Our bodies depend upon three main carriers of oxygen from the outside air into the cell processes. Oxygen gas is not very soluble in water and equally poorly soluble in the blood and body fluids. For example, one litre of blood only carries 2.7 ml of oxygen. This would be totally inadequate to support life. Arterial blood carries 2.7 ml O_2 per litre of blood dissolved in it and 204.7 ml chemically combined with haemoglobin.

The majority of the oxygen retained in our bodies is held by chemical bonding between the oxygen molecules and the iron atoms in myoglobin or haemoglobin. The myoglobin–oxygen compound acts as an oxygen store in muscle tissues. The blood haemoglobin is an efficient oxygen carrier to all parts of the body.

Each haemoglobin holds four oxygen molecules and picks these up in the oxygenation process in the lungs. The oxygenated blood circulates and at cells where the oxygen concentration (or partial pressure) is less than that of the haemoglobin–oxygen, the latter releases its oxygen. The deoxygenated blood continues around the circuit back to the lungs for further oxygenation. Any infection that prevents diffusion of the oxygen gas from the lungs into the blood will affect the efficiency of the whole body. Bronchitis, excess mucus, colds and influenza can all reduce the diffusion of oxygen into the blood and so result in shortness of breath when doing even simple movement or exertion. Smoking has a major effect on the efficiency of the lungs. Uptake of oxygen is also reduced at high altitudes.

The oxygen stored in muscle tissues is linked to myoglobin, with only one oxygen molecule for every myoglobin molecule. This compound holds onto its oxygen firmly and only releases it when the surrounding tissues have a large demand for oxygen, as in strong exercise. The oxyhaemoglobin then re-oxygenates it.

Factors that influence the release of oxygen from oxyhaemoglobin are partial pressure or demand for oxygen at the site (for metabolism), pH, carbon dioxide concentration at the site and body temperature.[3]

10.5.1 Some unusual bacteria

There are certain bacteria that like an oxygen rich atmosphere and others do not. Cohen, in *New Scientist*,[3] summarized research that showed that some 'flesh eating' bacteria, when injected into some cancers, consumed the dangerous cells and rendered them inactive. The critical thing is to choose bacteria that consume cancer cells in the absence of air inside the tumours but die as they reach the more oxygen-rich outside. This is even more effective if used in combination with chemotherapy. These treatments have worked in animals but it will be some time before they will be tested on human tumours.

Answers to the diagnostic test

1. Process of gain of oxygen by an atom or molecule (1)

2. A process of electron gain (1)

3. Hydrogen has been oxidized to water. (2)
   Copper oxide has been reduced to copper (2)

4. Super oxide dismutase (1)

5. Nitric oxide (1)

6. To thin the blood and take up any harmful free radicals (2)

Further questions

1. Look at the following chemical reactions and say what has been oxidized and what has been reduced:

$$2PbO + C \rightarrow 2Pb + CO_2$$
$$Mg + 2H^+ \rightarrow Mg^{2+} + H_2$$
$$C_{12}H_{22}O_{11} + 12O_2 \rightarrow 11H_2O + 12CO_2$$

2. $H_2S + O_2 \rightarrow SO_2 + H_2O$
   Balance the equation and explain what has been oxidized and what has been reduced.

3. Sometimes, in the oxidation processes of the body, instead of carbohydrates being oxidized to carbon dioxide and water, side reactions occur. In these hydrogen peroxide is formed in very small quantities and, if insufficient oxygen is present, small quantities of carbon monoxide. How does the body deal with these unwanted by-products?

4. It has recently been discovered that nitric oxide (NO, nitrogen II oxide to be precise) is present in the body. It performs many functions, even though it is only present in very small quantities and sometimes lasts only a fraction of a second. Sometimes it is stabilized by reacting with other substances. How could this compound have been made in the body? Useful web sites to research on this topic are: www.hhmi.org/science/cellbio/stamler.htm and girch2.med.uth.tmc.edu/faculty/fmurad/index.cfm

5. In an earlier chapter it was found that an oxidation product of making energy from carbohydrates was not only carbon dioxide and water but also a little carbon monoxide. Outline the advantages and disadvantages of having such a compound 'loose' in the body.

6. The most convenient remedy to take straight after a heart attack is either an artificial SOD agent or the usual 'man in the street' remedy of two aspirins. Explain the roles of these two chemicals in treating patients after a heart attack. Why are people who are at risk of heart attacks told to take half an aspirin a day?

References

1. J. Mann. Medicine advances. *Chemistry* 2000, 13–15 (special issue).
2. G. Thomas. *Medicinal Chemistry*, Chap. 11.5.1, p. 435ff. Wiley, Chichester, 2000.
3. T. Hales. *Exercise Physiology*. Wiley, Chichester, 2003.
4. P. Cohen. An appetite for cancer. *New Scientist*, 1 December 2001, 14.

Also see the articles: J. S. Stamler, Nitric oxide in biology at www.hhmi.org/science/cellbio/stamler.htm and L. J. Ignarro, Regulation and modulation of NO production at ignarro@pharm.medsch.ucla.edu

11 Analytical Techniques

Learning objectives

- To give an overview of some of the most useful analytical chemical techniques used in medical science.

- To show the relevance of chemical analysis to medical science.

Diagnostic test

Try this short test. If you score more than 80 % you can use the chapter as a revision of your knowledge. If you score less than 80 % you probably need to work through the text and test yourself again at the end using the same test. If you still score less than 80 % then come back to the chapter after a few days and read again.

1 What property of an atom or ion does mass spectroscopy use in analysis? (1)

2. What is meant by R_f in paper chromatography? (1)

3. What property of a compound does infrared spectroscopy use in analysis? (1)

4. What is the advantage of electron microscopes over conventional light microscopes? (1)

Chemistry: An Introduction for Medical and Health Sciences, A. Jones
© 2005 John Wiley & Sons, Ltd

5 What does MRI stand for?   (1)

6. What does PET stand for?   (1)

7. Which analytical method could be used for:

    i. the separation of a mixture of amino acids;   (1)
    ii. the location of a brain tumour;   (1)
    iii. the viewing of a sample of a cell abnormality;   (1)
    iv. the analysis of the chemical contents of a tube thought to contain
       traces of a glue-like prohibited substance?   (1)

Total 10 (80% = 8)
Answers at the end of the chapter.

Grandmothers watching young mothers protecting their offspring from picking up and tasting dirt have often been heard to say 'let them eat it as they have got to eat their peck'. This 'peck' often contains micronutrients only detectable once sophisticated analytical techniques were devised.

It is only the ability of modern chemistry to detect very small quantities of materials that made the following discovery possible. In some recent research it was reported that one subtle way in which cancer tumours cells differ from normal cells is how they metabolize carbohydrates present on their surfaces. Cancer cells have far more of the carbohydrate sialic acid, which can be detected with MRI (magnetic resonance imaging) analytical techniques. It was found that the sialic acid normally appears on the surface of the cells only in foetal development, but it appears abnormally in patients with gastric, colon, pancreatic, liver, lung, prostate and breast cancers, as well as in leukaemia. Research is continuing.[1]

11.1   The need for analysis

The modern world is in an active state of combat against disease, but it has been this way since time began. There were always witch doctors and old wives' tales of remedies for various ailments. It has only been in recent years that some degree of understanding of the cause and effect of certain diseases has been understood. Often minute quantities of a specific chemical can produce a cure or relief from pain for a particular complaint, but to discover and detect the exact chemical in a plant requires sophisticated analytical techniques.

Prince Albert died in 1861 at the age of 42, because the doctors at that time did not know how to treat typhoid fever (antibiotics had not then been invented). Albert was comparatively old when he died, as the average life expectancy in the early nineteenth century was between 40 and 50, and infant mortality was almost 20 %.

Thousands of people in Britain, and Europe died of cholera during the nineteenth century. At that time there was no known cure and no connection was made between the incidence of cholera and the unsanitary conditions. In the cities human excrement was thrown into open sewers flowing down the streets and into rivers. The rivers were also the source of drinking water. London was known as the city of smells. The disease was thought to be carried by the 'smells'. The conditions were only improved when major underground sewers were built in the 1880s. It was also realized that there were very small organisms such as bacteria and viruses. The bacterium responsible for cholera was discovered by Robert Koch in 1883.

The discovery of penicillin by Fleming in 1928, and many other antibiotics in the 1940–1950s and onwards, considerably reduced death rates from bacterial infections. Now the average life expectancy is almost 70 and increasing.[2]

From the 1950s onwards, the research and screening of newly discovered drugs led to a major breakthrough in knowing what molecules to look out for as possible remedies for complaints. Streptomycin was designed after considering the structures of other possible active molecules, and it was found to be particularly effective against tuberculosis and pneumonia.

The modern day scourge of the world is the rampant advance of HIV and AIDS, and many laboratories are working flat out to find a vaccine or a means to prevent the spread of the disease. The obvious way is protected and responsible sex, but people are not easily persuaded of this, particularly in poorer countries.

All the above-mentioned areas of research have only become possible due to the wonderful and creative application of analytical chemical techniques and purification procedures. Such procedures and apparatus are needed to detect small quantities of materials. They are also necessary for monitoring drug administration and effectiveness, and for analysis of breakdown products made in the body.

11.1.1 What techniques?

What follows is an overall and simplified view of some of the most commonly used analytical chemical procedures. Analysis is usually done with very small amounts of the samples, so the techniques must be very sensitive. The concentration of hormones in our blood is similar to that produced by adding a few grains of salt to a swimming pool. The methods generally depend upon the physical properties of atoms, molecules or parts of them. Some methods depend upon the interaction of the atoms and molecules with electromagnetic radiation (i.e. from very short-wave-length gamma rays through visible light to the long waves of the radio waves; Table 11.1).

Table 11.1   The range of electromagnetic radiation

	Gamma ray	X-rays	UV	Visible	IR	Microwave		Radio
Wavelength in m^{-1}	10^{-12}	10^{-10}	10^{-8}	Blue–red		10^{-4}	10^{-2}	10
	High energy					Low energy		

Some further methods depend upon the accurate measurement of mass. Others depend upon the 'stickability' of molecules to various surfaces (chromatography) and yet others upon their ability to take up the energy from various parts of the radiant energy spectrum.

It must be remembered that the materials that are being considered were not even suspected to exist at the beginning of the twentieth century. Once it was realized that microscopic quantities of materials can greatly affect our metabolic systems, these techniques became crucial in the development and understanding of the mechanism of drugs and the analysis of residues in urine, etc.

11.2   Mass spectroscopy

This technique uses the fact that particles of different masses will respond differently to electrical and magnetic forces. The lighter particles move more easily than the heavier ones.

A very small quantity (as little as a microgram, 10^{-6} g) of a sample material to be analysed is heated to convert it to its vapour. It is then subjected to a stream of fast moving electrons. This has the effect of charging the fragments of materials present in the sample, forming ions:

$$X \text{ (gas)} + e^- \rightarrow X^+ \text{ (gas)} + e^- \text{ (slower moving)} + e^- \text{ (from the sample)}$$

If the sample is of an organic origin (including human samples), the fast moving electrons can cause the molecules present to fragment at various places in their structure, thus forming a wide range of ions. These fragments will all be characteristic of the original molecules present and be a bit like a fingerprint of the molecule.

These ions are accelerated by a strong electric field and focused into a beam. The beam passes through a strong magnetic field. All the particles in the beam are affected in the same way because they all have a charge (positive or negative). The less massive particles are affected more. Think of it this way: you are driving over the Severn Bridge. You are driving a white van fully loaded with gold ingots (you are a notorious bank robber). Ahead of you is an exactly similar van carrying a load of feather pillows. Suddenly a side wind starts to blow. Both vans swerve. Which one swerves more? A similar thing happens to the ionized particles. The lighter particles deviate from their paths more than the heavier ones.

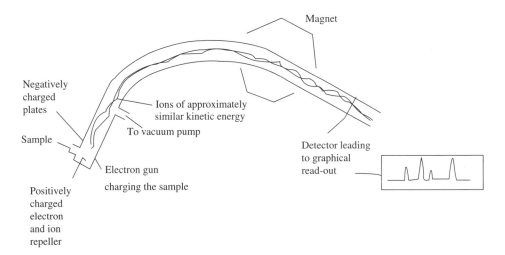

Figure 11.1 Mass spectroscopy

As the particles arrive in a detector, their charges and quantities are recorded (Figure 11.1). The pattern of the read-out chart shows up the masses of the fragments that are present. Then a trained technician can work out where the particles came from. By this method the drugs and metabolites can be characterized. This separation technique depends upon the different kinetic energies ($1/2mv^2$; m = mass, v = velocity or speed of the particles) of the particles. The pattern of the mass spectrum is characteristic of the materials it comes from. The height of the graph tells you the amount present, and the position of the peak from the starting point tells mass–charge ratio and size of the fragments (Figure 11.2).

This technique is about 1000 times more sensitive than infrared and magnetic resonance spectroscopy, outlined later. It can also be used to detect different masses of radioactive and nonradioactive isotopes of elements. It can distinguish between the carbon 12 and 13 isotopes and also the oxygen 16 and 18 isotopes. By

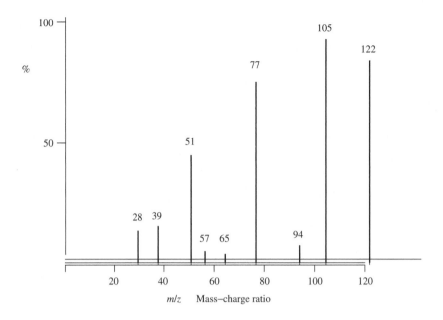

m/z	Ion fragment		
122	M$^+$	Structure	
105	M–OH$^+$	Structure	
77	M–OH–CO$^+$	C$_6$H$_5^+$	
51	M–OH–CO–C$_2$H$_2^+$	C$_4$H$_3^+$	

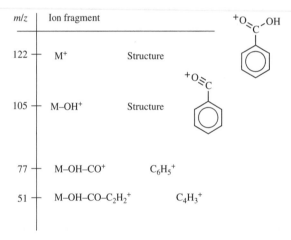

Figure 11.2 Mass spectrum of benzoic acid showing major fragments

incorporating these isotopes into a synthetic drug, the technique can be used to trace and monitor the pathway of labelled drugs through a human system.

Mass spectroscopy has been extensively used in the following situations:

- analysis of urine samples from athletes suspected of drug abuse (e.g. in the Olympic games 2000 and 2004, when detection of the use of nandrolone, testosterone and other substances occurred);

- detection of the presence of LSD in drug abuse cases by the fragments of the metabolites in urine;

- analysis of small samples of baby urine when investigating SIDS (sudden infant death syndrome);

- testing of the purity of foodstuffs, and safety of additives;

- analysis of prostaglandin in human semen.

11.3 Chromatography

The techniques of chromatography are used mainly for the separation of materials that are normally difficult to separate and are present in very small quantities in mixtures. There are various methods of chromatography, but the principles are similar. They all depend upon a material for the sample to 'stick' to and a solvent (liquid or gas) to push the sample along. The absorbent surface can be paper or a thin layer of a suitable unreactive powder, like alumina. This is called the stationary phase. In thin layer chromatography (TLC), the material can be chosen for ease of separation.

In paper chromatography the stationary phase is the cellulose of the paper, and the spot of sample is added to the paper. A suitable solvent, say water or alcohol, is called the mobile or moving phase. This is added to push the materials across the paper or a thin layer of powder as it soaks across. Each material in the mixture to be separated will have a different solubility in the solvent and also a different degree of adsorption (sticking) to the paper. As the solvent soaks along the paper, the sample repeatedly becomes dissolved in the solvent and adsorbed and desorbed to and from the paper. These variables make the time it takes for a material to move along unique for each material in the mixture.

The distance substances a and b in Figure 11.3 moves compared with how far the solvent, x, moves, i.e. a/x and b/x, is called the R_f value. The R_f values are characteristic for the individual substances using the same type of paper (or thin layer of inert material), same solvent and same method. These values are standard, whether the experiment is repeated in Tokyo or UK, as long as the same conditions are applied.

Gas chromatography can also be used (Figure 11.4) to separate mixtures of gases or vapours using a heated column of powder. Detection of the arrival of the components of the mixture at the end of the column depends upon their properties,

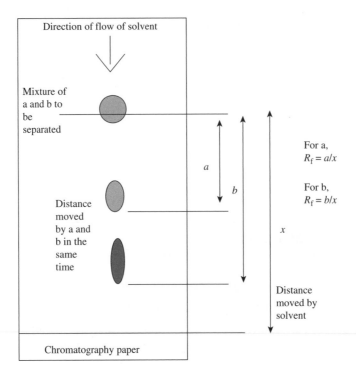

Figure 11.3   Paper chromatography

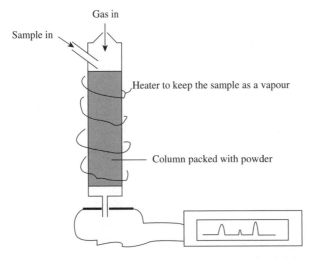

Figure 11.4   Gas chromatography

e.g. colour, conductivity or by passing the gases through a small flame and watching (electronically) how the flame changes colour as each substance arrives. Alternatively a series of samples collected throughout the analysis can be subjected to infrared, magnetic resonance spectroscopy or mass spectrometry. Any mixture that can be vaporized can be separated by this method. The column can be kept warm to maintain the substances as vapours.

Whichever technique is used – paper chromatography, thin layer chromatography, a column of material – separation will be achieved, but different rates of flow or R_f value will be recorded. Different powders are used to achieve different separations. Research in the area of analytical chemistry means that different variations of the techniques are being developed periodically. The products can be separated using chromatography then the separated components characterized using mass spectrometry or various types of spectroscopy.

11.3.1   Nobel prize

In 1962, the Nobel Prize winners for Physiology and Medicine were Francis Crick, James Watson and Maurice Wilkins. They used chromatography to separate the complex mixture of amino acids making up proteins. This led to the characterization of the structure of a protein by X-ray analysis and in particular the realization that the three-dimensional structure of DNA was an inter-linked double helix.

11.3.2   Applications

Chromatography, in its various forms, is still extensively used in medical pathological laboratories, both in screening natural products for their components and in analysis of body samples. Internationally, the Jockey Club, along with the authorities heading international athletics, takes random samples of urine. This is for analysis for the presence of illegally used drugs that could enhance performance, i.e. dope testing. The officials at the 2004 Olympic games had the most sophisticated analytical chemical machines at their disposal. Some athletes were probably using drugs which are not yet easily detectable by their metabolites (break-down products). Also, some athletes claim that they have not taken drugs and that their own metabolism is producing metabolites similar to those coming from banned drugs like 'nandrolone'. Who knows, in the future it might be possible to use gene technology and therapy to develop naturally body-building materials without taking any drugs at all, but will these 'natural' products be detected or even found acceptable to the sports authorities? Is this cheating?

Sports medicine is a growing industry, and so are the analytical methods required to detect materials and drugs used both legally and illegally. Of the 10 000 athletes taking part from 200 countries in events like the Olympics, some will be using artificial means to survive the gruelling sporting tasks, but the ancient Greeks also used 'natural' products to enhance their performance. Seeds, oils, diets, mushrooms and even ground up asses' hooves were all reported to have been used in the second century games.[3]

11.4 Spectroscopy of various types

11.4.1 Infrared spectroscopy

Infrared spectroscopy is mostly used to detect the functional groups of atoms present in a sample, for example, OH groups or $C=O$ groups in organic molecules. It is mostly used for organic compounds. The height of the detected 'peaks' on a graph is proportional to the amount of that particular group in the materials. The groups in a molecule that are activated when exposed to infrared radiation are those that are unsymmetrical, and cause the groups to stretch, flap or move and twist in such a way that their centre of gravity is displaced (Figure 11.5).

Samples used for infrared analysis must be free from water because it absorbs so much radiation itself that it obliterates much information from other groups that might be present. Therefore samples of a material are dried and usually mixed with Nujol oil, or compressed with potassium bromide powder and put under great

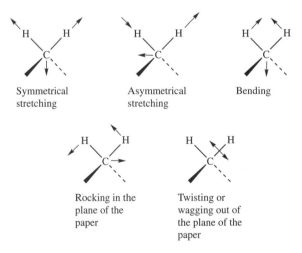

Figure 11.5   Effect of infrared radiation on molecules

pressure to form a thin glass-like transparent disc. The samples are then put in a beam of IR light. The resulting light, after passing through the sample, is compared with the original IR light. The difference between these values shows what has been absorbed by the sample. The bonds holding each group of atoms together have their own characteristic amounts of energy that make them move, stretch, twist, etc. Thus, when that exact wavelength is absorbed, you know that those bonds are present. This technique is particularly useful for molecules in samples that are of organic origin, things like lipids, proteins and smaller molecules.

Drunk drivers know about the accuracy of IR analysis for alcohol in the urine, blood or breath. Alcohol goes into the stomach and is absorbed into the blood and transferred to the liver. Here the alcohol is metabolized and removed at a slow rate. The blood also transfers the alcohol to the heart and lungs where it is 'oxygenated'. Some then passes into the breath. The amount of the alcohol in the breath is directly related to the alcohol in the blood and stomach. They are in the ratio 2300:1, so it is easy to calculate how much alcohol has been consumed by analysing the quantities of alcohol in the breath. The alcohol present is detected by the characteristic shape of the graph from infrared spectroscopy. Infrared analysis will also detect ethylethanoate in the breath of people who have been glue sniffing since ethylethanoate, $CH_3COOC_2H_5$, contains C—H, C—O, C—C bonds (Figure 11.6).

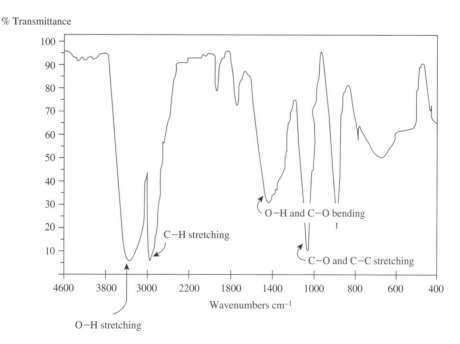

Figure 11.6 Infrared spectrograph of ethylethanoate

11.4.2 Ultraviolet and visible spectroscopy

These techniques depend upon the energy taken up by molecules from visible light and ultraviolet light. These techniques can detect very small changes in wavelengths. This shows up as a change in colour in visible light or a shift of energy for the UV spectrum.

Many parts of organic molecules absorb light and so cause colour changes and wavelength shifts. The presence of certain groups in a molecule can be shown by reacting them with certain dyes. The intensity of the colour or absorption of the light is directly proportional to the amount of the materials present.

11.5 Electron microscopes and scanning electron microscopy (SEM)

When looking at things with an ordinary light microscope, there is a limit below which very small items cannot be seen. Using a beam of electrons instead of light can often show up extremely small things (Figure 11.7). It cannot be used to obtain images of living things as the samples have to be set in a solid material and must be studied at very low pressures. The dead remains of a living system (plant and animal tissue) can all be 'seen' using an electron microscope. The procedure is quite complicated and not as easy as using an ordinary microscope, but the resolution is much greater and able to achieve magnifications of 1×10^6. The procedures are usually used for looking at the surfaces or sections of materials, including the surfaces of bacteria. This helps to show how they attack human cells. It is not yet possible to see individual atoms or very small molecules.

When an incident beam of electrons is focused on a specimen at low pressures then many things can happen. Some electrons hit the nucleus of atoms and are bounced and scattered back towards the source. The heavier the atoms, the more back-scattering occurs. Some of the electrons cause electrons in the sample to be ejected. Some of the incident electrons knock out secondary electrons from the sample and the energy of these is characteristic of the atoms concerned. Some electrons cause X-rays to be emitted from the sample and some electrons pass straight through the sample.

The technique is called a scanning process because the beam of electrons are carefully directed across the sample until all the sample has been scanned. The analysis of all the radiation emitted is computed and a two-dimensional image is created either on a screen like a television screen or on a photographic plate. One disadvantage of this technique is that for nonconductors of electricity the sample has

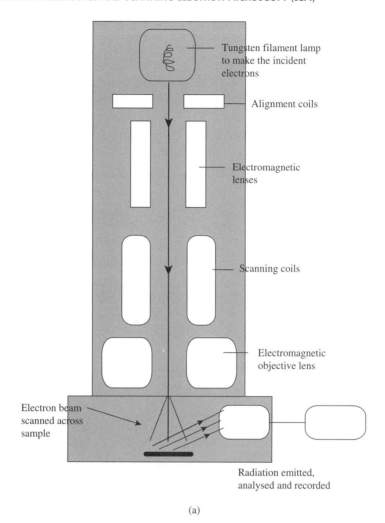

(a)

Figure 11.7 (a) Scanning electron microscope. (b) Diagram of the types of electrons and radiation that are scattered from the atoms of a sample in SEM. (c) Section of a rat liver at ×30 000 (*continued overleaf*)

to be very thinly coated with a layer of gold. Any scattering from the gold can be allowed for when analysing the data.

SEM is particularly useful for showing up the surface structure of materials by analysing the secondary electrons. Transmission electron microscopy (TEM) relies on the use of the electrons passing through the very thin samples and can show up images of the internal structure of samples. It can achieve a resolution of about 1×10^{-10} m. Both SEM and TEM require a high vacuum and so samples must be stable in vacuums and when subjected to fast moving electrons.

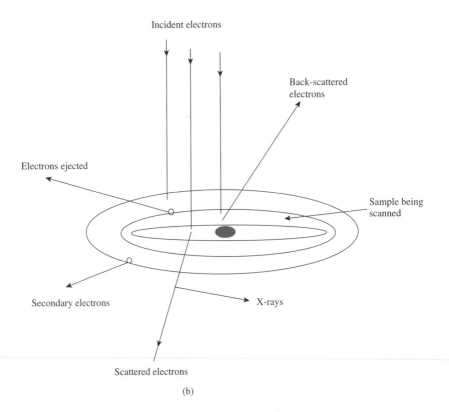

Incident electrons

Back-scattered
electrons

Electrons ejected

Sample being
scanned

Secondary electrons

X-rays

Scattered electrons

(b)

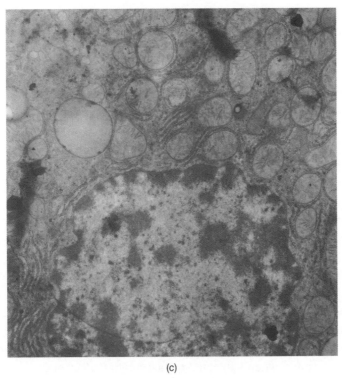

(c)

Figure 11.7 (*continued*)

11.6 Magnetic resonance spectroscopy (MRS) or magnetic resonance imaging (MRI)

The original name for these techniques was nuclear magnetic resonance (NMR), but the word 'nuclear' has been dropped in medicine to avoid any association with radioactive radiation. This technique gives information about the environment in which the nuclei of atoms find themselves in molecules of compounds. It can be used to give information about individual molecules as well as producing images of soft biological tissues.

The nucleus of an atom in a molecule spins in the presence of an external magnetic field. It can align its own spin either in the same direction as an external applied magnetic field (+), or opposite to it (−). The collection of effects that these spins have characterizes the atoms present in a molecule.

The spinning and oscillation of the atoms are initially set off using a burst of radio waves. The magnetic effects that this oscillation causes are then analysed. The values from hydrogen atoms in different environments are the set of values that give most information. So any soft tissue, organs or limbs that contains water can be viewed using this technique. It is used to measure the effects of materials in the body. These can be observed as they are proceeding because MRS is not harmful to tissues. There are no known side effects of magnetic resonance and so even the young and the elderly can be scanned frequently without any bad effects. The process does, however, require the patient to remain still inside a large tube

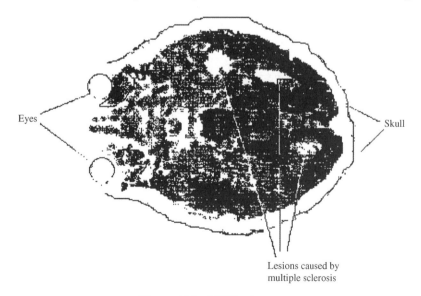

Eyes

Skull

Lesions caused by
multiple sclerosis

Figure 11.8 MRI brain scan

surrounded by a very strong magnet. A body scan can reveal different densities of materials and so show up where abnormalities are occurring within any of the body organs. The images that are obtained look a bit like X-ray photographs and are called MRI or magnetic resonance images.

The technique is widely used in the diagnosis of cancers, brain tumours, hydrocephalus and multiple sclerosis. It was reported in *New Scientist*[4] that very small (4 mm) breast cancers can be detected by measuring the diffraction pattern given off by them. Figure 11.8 is a simulation of the type of image obtained from a MRI scan for a patient with multiple sclerosis.

Because of the very strong magnetic fields used in this technique, all iron and magnetic material must be excluded or at least screwed to the bench. Otherwise they can be attracted very powerfully and things can fly like missiles towards the magnets. All instruments must be nonmagnetic. It is not advisible to wear an expensive watch near these machines!

11.7   General conclusions

In any form of detective work or diagnosis no single test result is conclusive by itself, but has to be confirmed by at least one other method. The same is true with what you have just read about the methods used for analysis of materials. Each method brings to light a new aspect of the properties of the substance. When all the data is collected together it provides an informed, highly probable, scientifically backed 'guess'.

Analytical methods involving radioactivity are discussed in Chapter 12, including PET (proton emission topography).

Perhaps you can see the reasons why some of the samples sent to the laboratory for analysis can take some time to process, as none of these techniques are instant and they all require time and expertise. If you have an opportunity, follow the pathway of a sample you send to the laboratory for analysis and watch the procedures being used. It could be part of your continuing professional development plan.

Answers to the diagnostic test

1. Its mass   (1)

2. Distance moved by the sample divided by distance moved by solvent
   in paper chromatography   (1)

---

3. Flapping of a molecule and particularly those movements that
   result in a displacement of the centre of gravity of the molecule   (1)

4. It uses electron beams to view the sample which is shorter
   wavelength than a light beam and so can see smaller objects   (1)

5 Magnetic resonance imaging   (1)

6. Proton emission topography   (1)

7. i.   Chromatography   (1)
   ii. MRI or PET   (1)
   iii. Electron microscopy   (1)
   iv. IR spectroscopy   (1)

Further questions

1. i.   How would you separate the components of a sample of an aqueous solution
        thought to contain a sugar, glucose, an amino acid, glycine and water by paper
        chromatography?
   ii. How could the components be characterized to know which is which?
   iii. Write down the structure of glycine and glucose.
   iv. What are the characteristic parts of each of the two components. Would they be
        expected to differ in their IR values? Explain your reasoning.
   v.   Can these two molecules (from iii) be seen using an electron microscope?
   vi. If you had another unknown sample thought to contain glycine, how could you
        confirm it?

2. If an athlete was thought to be unfairly taking body-building drugs, what methods
   could be used in their detection?

3. Which of the following techniques can be used to follow a mechanism in a living
   system and not harm the patient: mass spectra; IR spectroscopy; chromatography;
   nuclear magnetic resonance and MRI; electron microscopy?

4. 'The advances in analytical chemistry have helped in the war against disease.' Discuss
   this statement.

References

1. D. Bradley. Seek and ye shall find. *Chemistry in Britain*, July 1999, 16.
2. S. Aldridge. A landmark discovery. *Chemistry in Britain*, January 2000, 32–34.
3. B. Kingston. Catching the drug runner. *Chemistry in Britain*, September 2000, 26ff.
4. D. Graham-Rowe. X ray trick picks out tiny tumours. *New Scientist*, 22 February 2003, 14.

In addition, a useful reference is *Modern Chemical Techniques* compiled by B. Faust, Royal Society of Chemistry, London. The Royal Society of Chemistry produces a CD entitled *Spectroscopy for Schools and Colleges*. This contains examples of the types of spectra obtainable from analysis of various molecules. It also has video clips of how spectra are run.

12 Radioactivity

Learning objectives

- To give a basic understanding of the principles of radioactivity.

- To show the nature and uses of radioactivity.

- To outline the different types of radiation emitted from some isotopes.

- To see the application of radioactivity to medicine.

Diagnostic test

Try this short test. If you score more than 80 % you can use the chapter as a revision of your knowledge. If you score less than 80 % you probably need to work through the text and test yourself again at the end using the same test. If you still score less than 80 % then come back to the chapter after a few days and read again.

1. What is meant by:

   i. Atomic number   (1)
   ii. Proton and neutron   (1)
  iii. Isotope   (1)
   iv. Alpha particles, α (1)
    v. Beta particles, β (1)
   vi. Gamma radiation, γ (1)

Chemistry: An Introduction for Medical and Health Sciences, A. Jones
© 2005 John Wiley & Sons, Ltd

2. Explain how the three main types of radiation, alpha, beta and gamma, differ in their properties. (5)

3. Explain what is meant by the 'half-life' of a radioactive element and its significance in dating rocks or ancient human remains. (5)

4. Give some positive arguments for the use of radioactive isotopes in medicine. (4)

Total 20 (80 % = 16)
Answers at the end of the chapter.

What does radioactivity have to do with radios? Where in a house would you find the radioactive element americium? The general public seems to be fearful of anything to with radiation. We are surrounded by helpful and harmful radiation. Some radiation can be manipulated for our special use, including microwaves for cell telephones and at a different wavelength for cooking. We are shielded from the radiation coming from space by the ozone layers in the upper atmosphere. Some wavelengths of radiation get through and give us sunlight and heat. Artificial radiation from disintegration from a range of nuclei of various isotopes can be manipulated to attack harmful cancers, as can laser devices.

12.1   Introduction to the effects of radiation

It is perhaps worthwhile to summarize what an atom looks like. It contains very heavy protons and neutrons in its nucleus and this is surrounded by shells of spinning electrons. This chapter generally has to do with the nuclei of the atoms. The electrons are still there, but it is the nucleus that interests us.

The nuclei of elements of low atomic number have roughly the same number of neutrons as there are protons, e.g. helium has two protons and two neutrons. Some elements have a series of variations of number of neutrons for the same atomic number (number of protons). Elements that have the same number of protons but

different numbers of neutrons are said to be isotopes of each other. Some of these combinations are stable arrangements, but some are distinctly unstable and their nucleus breaks up. These arrangements are said to be unstable isotopes. This tendency occurs more often in the very heavy elements like uranium. Uranium has an atomic number of 92 but it can have over 140 neutrons in its nucleus. This causes the nucleus to be unstable and gradually break up, giving off particles in order to try to stabilize itself. The throwing out of these particles or radiation is called 'radioactivity'.

When you are hit by a hard, fast-moving cricket ball, it hurts. The energy of the ball is transferred to you and might leave a bruise where the ball has damaged some of the skin, blood and flesh cells. High-energy radiation comprises much smaller particles and they can also damage the cells.

Radioactive elements have unstable nuclei that give out energetic particles and gamma rays and can damage body cells close to the source. The alpha and beta particles are very small, so damage to the body cells only occurs if they are exposed for a long time or if the source is close by and very energetic. Most of the radiation from radioactive sources is stopped by air, material clothing, building material or other protection.

We are exposed to many types of natural radiation from space, rocks of the environment, X-rays, radiotherapy, or even by sitting too near a TV set. All of these are usually not large enough to cause any adverse effects to our overall body health.

People working with high-energy sources of radiation are tested regularly, take special precautions and wear protective clothing. Airline pilots and crew are exposed to higher than normal amounts of cosmic rays and are regularly tested for any cell damage.

12.2 Isotopes and radioactivity

Each element has a unique *atomic number*. The atomic number of an element tells you how many protons it has in its nucleus. The atomic number of chlorine, for example, is 17: this tells you that every atom of chlorine has 17 protons in its nucleus. Since there are 17 positive protons in the nucleus, there are 17 negative electrons orbiting the nucleus. The atom is electrically neutral. The number of protons never changes. No matter where a chlorine atom is found, or where it has come from, or what it is doing, or if it is in a compound with other elements, it will always have 17 protons in its nucleus.

The nucleus of an atom also contains neutrons. Neutrons are neutral – they have no electrical charge. Therefore, in theory, if an atom of chlorine had 17 protons and one neutron in its nucleus it would make no difference to its chemical properties. That is because it is the number of electrons in the outer electron shell that controls chemical properties, and it is the number of protons in the nucleus (the atomic number) that characterizes what the element is. Therefore, in theory, the nucleus of an atom of chlorine could have any number of neutrons in its nucleus – it would still be an atom of chlorine as long as it has 17 protons. So what is the difference between these atoms with varying numbers of neutrons? There are two answers (aren't there always?).

Answer number 1 has to do with *mass*. This is the easiest answer. Electrons have almost no mass at all. They are so small that, for all practical purposes, they have a negligible mass. An electron has roughly one two-thousandth of the mass of a proton. Protons and neutrons have almost exactly equal masses. So if we call the mass of a proton or a neutron '1' – there are no units, no grams or pounds or ounces, because these particles are the basic units of matter – we say that the *mass number* of a proton or a neutron is 1. Consequently, we also say that the mass number of an electron is 0 (well, almost zero).

Isotopes of an element are atoms that have the same atomic number, but different mass numbers. All atoms of the same element have the same atomic number. Different atoms of the same element might have different numbers of neutrons, i.e. different mass numbers.

Answer no. 2 is to do with *stability*. As it happens, if you take a stroll around the universe and examine every chlorine atom you ever come across, you will find none that contain no neutrons. They will all contain 17 protons, and they will all contain some neutrons. In fact, you will find that almost all the chlorine atoms you find will contain either 18 or 20 neutrons. Why?

There seems to be some kind of relationship between the number of protons and neutrons in a nucleus. For chlorine, 17 protons to 18 neutrons seems to be good – it is a stable relationship. In addition, 17 protons to 20 neutrons also seems to be good also for chlorine – it is also a stable relationship. In other words, all the chlorine atoms you might find will have an atomic number of 17, and mass numbers of either 35 or 37. The mixtures of these naturally occurring chlorine atoms in the gas are in such a ratio that the average atomic mass of normal chlorine is 35.5 (Figure 12.1).

What happened to all the other possible combinations of protons and neutrons that might have existed when the universe began? Only certain combinations of neutrons and protons form stable arrangements. The 35 and 37 forms of chlorine do not split up. These nuclei are not unstable and do not disintegrate. They are not radioactive.

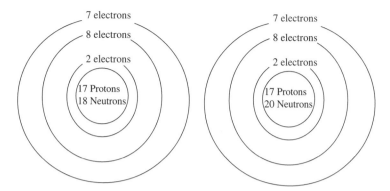

Figure 12.1   Isotopes of chlorine

Every element has isotopes. Most are stable and do not break up. Some unstable isotopes of the heavier elements are called 'radioactive', because they give off radiation when they split up.

12.3   Splitting the nuclei of atoms

When the nucleus of an atom is unstable, because the ratio of protons to neutrons is 'wrong', the nucleus splits up to try to obtain a stable arrangement (Figure 12.2). This breaking up is called 'fission'. Nobody can tell exactly when any single atom will break up, but if you take a real sample of an element it will contain billions of atoms and we can start to make predictions. We do not know what is going to happen to an individual atom, but we can say that, if we take several billions of atoms, half of them will have split up (or 'decayed') after a particular length of time. This is

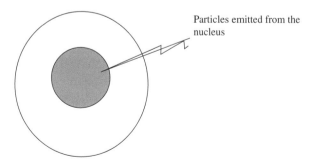

Figure 12.2   Particles emitted from an unstable nucleus

known as the *half-life* of the particular isotope of the element. The half-lives of the least radioactive isotopes are very long – it takes a long time for the number of atoms to break up and leave half that original number. The most radioactive elements have short half-lives. When the nucleus splits up, any one or more of three types of radiation are given out – alpha, beta and gamma.

12.4   Properties of alpha, beta and gamma radiation

These different types of radiation have different properties.

12.4.1   Alpha particles

Alpha radiation is made up of quite large packets of particles, two neutrons and two protons, but no electrons. It consequently has two positive charges associated with it. In fact it is similar to the nucleus of a helium atom and so can be represented as $_2^4\text{He}^{2+}$. Because it is quite large, it does not move too quickly or very far when it spurts out of the nucleus of a radioactive atom. In its short lifetime it picks up a few electrons and becomes helium gas. It usually picks up these electrons from the air or material in which it is packed.

$$_2^4\text{He}^{2+} + 2\text{e}^- \rightarrow \text{He}$$

However, within these few centimetres it can be quite dangerous to body tissues. It is typically stopped by about 2 cm of air. It is stopped by a few sheets of writing paper, and certainly stopped by a thin sheet of metal. Because they have a mass of 4 and some movement energy, alpha particles can do considerable damage to body cells if they are in close proximity, and particularly if radioactive material is ingested.

In the 1930s, the alpha-emitting element radium was used in making 'luminous' paint for watch and clock faces. The alpha particles from radium interacted with the other contents of the paint and gave off light energy. The numbers glowed in the dark. It was safe for the watch wearers since the metal back of the watch protected the wrist from radiation damage. Nobody ever put their eyes close enough to the watch face to damage their eyes. However, the people who painted the numbers onto the watch faces licked their brushes to get a fine point. These people suffered horrendous mouth cancers and stomach cancers and many died.

Alpha particles (and also beta and gamma rays) given off by natural rocks around us will generally not damage us, as surrounding materials and air render them harmless. This process is called 'natural radioactivity' or 'background radioactivity', and some rocky areas of Britain have high (but not dangerous) levels of background radiation.

When an alpha particle is given off from the nucleus of an atom such as radium, that particular atom changes to an element two places to the left in the periodic table. The loss of two protons from an atom means it becomes a new element with an atomic number 2 units less. When a beta particle is lost, the mass is unchanged. There is a loss of a negative charge, but the proton on the nucleus remains so the positive charge goes up by one and the element moves one place to the right in the periodic table. If this process continues, it will eventually finish up as an element with the stable nuclear arrangement of a nonradioactive isotope of a different element. This element is usually no. 82, lead, with 125 neutrons. The slide down the periodic table is traced by a radioactive pathway is shown in Figure 12.3. Also

$$n \rightarrow p^+ + e^- \tag{12.1}$$

Each radioactive isotope has a similar breakdown pattern before it ends up some millions of years later as a stable, nonradioactive isotope, usually lead. There are a few different series like the one shown.

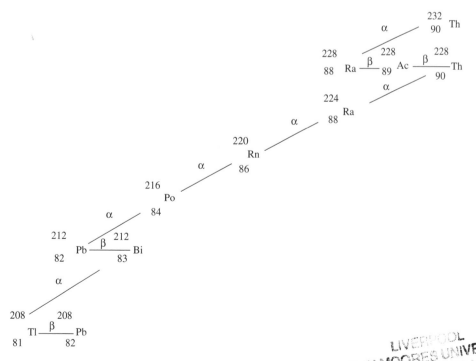

Figure 12.3 A radioactive decay series

12.4.2 Beta particles

Beta particles are very small. They are fast moving electrons that are shot out of the nucleus of a radioactive element. They are easily stopped by a few sheets of paper, a thin sheet of metal or about 20 cm of air. Even the most energetic beta particle can only travel a metre or so in air before it is stopped. It has a penetrating power 100 times greater than alpha particles. You do not keep a radioactive element giving off this radiation in your pocket!

Elements giving off this particle change to an atom one place to the right in the periodic table (as a negative charge is lost but the matched proton still remains, so the atomic number is increased by 1), see Figure 12.3.

Notice that these two processes change an element into a new one. The ancient alchemists tried all their lives to do this and change lead into gold. They did not succeed. They could not 'transmute' the elements. If they only knew it, nature was doing this all the time, although not making gold, as they wanted.

Uranium, $^{238}_{92}U$ loss of $^{4}_{2}He \rightarrow {}^{234}_{90}Th \rightarrow$ loss of two beta partiles $\rightarrow {}^{234}_{92}U$, etc.

12.4.3 Gamma radiation

When alpha and beta radiation are given off, the resulting nuclei shake down, like marbles in a bag, into some stable arrangement. Often gamma radiation is also given off in this process. Gamma radiation is made up of a high-energy photon with a very short wavelength and is similar to X-rays, but it is even more penetrating and damaging to body tissues. It has a penetrating power 10 000 times greater than that of an alpha particle. A thick block of lead is needed to stop this radiation, or even thicker walls of concrete. They are, however, beneficial if they are specifically targeted at cells needing to be killed off, e.g. cancer cells. Accurately calculated amounts of gamma radiation are used.

All three types of radiation are termed 'ionizing radiation' as they strip off electrons from atoms they collide with to leave a charged ion behind. Sometimes this property can be used to detect the presence of these rays. Geiger counters use this property. This radiation also 'fogs' a photographic film. This is shown up when it is developed.

Because two of these types of radiation (alpha and beta) are charged particles, they can be deflected by electrical and magnetic fields. This is not so for the gamma rays – they carry straight on . . . and on . . . and on, sometimes for miles. In the Hiroshima atomic bomb it was the high-energy gamma rays that vaporized people and vegetation. People further away suffered radiation burns from the gamma rays.

12.5 Half-life

The half-life measures the time it takes for the radiation coming off any mass of an isotope to drop to half its value. Short half-lives of hours or minutes mean that these isotopes give off radiation at such a rate that they deliver their radiation in a short time. Often, the shorter the half-life, the more energetic and more harmful the radiation.

Measurement of the amounts of radioactive isotope in a material can give a good indication of the age of the material. It can be used to determine the age of rocks and also organic materials in archaeological remains. It is often used in dating ancient plant or human remains by measuring the amount of radioactive carbon, C 14, still remaining in the samples. The method is called 'carbon dating'.

To ensure that people working with radioactive materials are protected from the harmful effects of radiation, they wear 'photographic' badges. These are developed periodically and the greater the exposure to radiation, the greater the 'fogging'. When the value becomes too high, the worker is taken off the work task.

Radiation is also detected using Geiger counters, which convert the incoming radiation to a click. The greater the intensity of the clicks, the greater amount of radioactive material present. These instruments always click in ordinary air because there is always some radiation entering a room from space, from the stone of the building or the rocks below.

There are limits to what is permitted for each user and this is governed by the Health and Safety regulations. The same applies to people working with X-rays. A good set of free wall charts is available from the National Radiological Protection Board (NRPB)[1] that summarizes all these processes.

The recognized unit of dosage is the 'sievert' (Sv). This unit measures the exposure to all forms of ionizing radiation, both radioactive and that from outer space, e.g. cosmic radiation. The average person in the UK receives about 2.5 mSv of radiation per year. There is a higher than average value in rocky areas of the UK, including Cornwall, with averages about 7.8 mSv overall per year. The average person working in the nuclear industry only receives about 4.5 mSv per year. All these values are considered 'low' and not a health hazard. Airline pilots receive from cosmic rays alone about 2.5 mSv per year and so their flying time needs to be monitored and managed correctly.

12.5.1 Isotopes in medicine

Isotopes of varying half-lives are used extensively in medicine both for diagnosis and also treatments (Table 12.1). A typical example of medical use of an isotope is

Table 12.1   The uses of radiation isotopes in medicine

Isotope	Half-life	Radiation	Uses
Chromium 51	28 days	Gamma	Diagnosis of spleen and gastrointestinal disorders
Cobalt 60	5 years	Beta and gamma	Cancer therapy treatment
Iodine 123	12 h	Gamma	Thyroid diagnosis
Iodine 131	8 days	Beta and gamma	Thyroid diagnosis and treatment
Iron 59	45 days	Beta and gamma	Bone, anaemia diagnosis and treatment
Krypton 81	2×10^5 years	Gamma	Diagnosis of lung and ventilation disorders
Technetium 99	6 h	Gamma	Diagnosis of disorders in heart, brain, liver and kidney
Thallium 201	7 h	Gamma	Investigations of coronary functions

the investigation for any dysfunction of the thyroid. Here a patient is given a drink of a suitably diluted solution of iodine 123, usually as the compound sodium iodide, NaI. The thyroid is then scanned externally for the arrival and disappearance of the gamma-emitting 123 iodide. An overactive thyroid will absorb more iodine then a normally acting one. A computer-generated image of the areas of the thyroid could reveal normal functioning or locate areas of abnormality, nodules or tumours. The short-lived 123 isotope quickly loses its radioactivity and leaves no aftereffects. Its half-life is about 13 h and it gives off gamma radiation. Because the thyroid absorbs any iodine, after diagnosis it is possible to treat the abnormalities with another more potent and longer lasting (less potent) radioactive isotope of iodine, namely iodine 131, which is a beta and gamma emitter. The radiation goes to the seat of the problem and is then able to attack directly any tumour cells present. The radiation is localized and does not affect any nearby organs. The use of radioisotopes is an expanding area of research. It has many applications to diagnosis, treatments and therapy. Other isotopes that are used to trace metabolic pathways in plants and animals are carbon 14, phosphorus 32, sulfur 35 and others.

12.6   Radiation everywhere

All the various forms of radiation cause damage to living tissue, but our universe is flooded with this sort of radiation – from the rocks under our feet, to the stones of our buildings to the cosmic rays from space. Fortunately most of the radiation is stopped by air, clothing or walls.

We can use the effects of radiation for good as well as evil. Radioactive isotopes of elements can be used in specific apparatus to focus their radiation on unwanted body material and cancerous cells. Often the gamma rays from the 60 isotope of cobalt ($_{27}^{60}$Co) are used to produce these penetrating rays to kill cancer cells. This same isotope is also used to produce the gamma rays to sterilize medical instruments. It kills the germs but the instruments remain unaffected.

Skin cancers can be treated and killed off using the less penetrating beta rays from either phosphorus $_{15}^{32}$P or strontium $_{38}^{90}$Sr. The exact amounts of radiation needed to perform these tasks must be accurately calculated.

We can also use radioactive isotopes to trace the course of events proceeding in the body. We can persuade our patient to eat, drink or to have injected a substance that contains a safe and easily eliminated radioactive isotope. Then we can use a radiation detector (photographic plate or Geiger counter) to follow the course of the radioactive isotope around their system. This can detect blockages and the flow rates of fluids, and trace various metabolic mechanisms. Technetium, $_{43}^{99}$Tc, is often used to detect brain tumours. A sodium isotope, $_{11}^{24}$Na, is used to follow the movement of sodium ions, Na$^+$, through the kidney. The very sensitive tracing of the blood flow in the brain often uses the $_{8}^{15}$O positron emitter (a positron is a particle the size of an electron, but with a positive charge). The technique is called PET or positron emission tomography. The normal safe nonradioactive isotope of oxygen is $_{8}^{16}$O.

12.6.1 Units of activity

The activity of a radioactive sample is quoted in 'Becquerels' (named after the discoverer of natural radioactivity) and it varies from one element to another. One atomic disintegration per second is called one Becquerel (Bq). A further unit often used is the Curie (Ci), which is the activity produced by 1 g of radium 226 and is equal to 3.7×10^{10} disintegrations per second.

$$1\text{Ci} = 3.7 \times 10^{10}\,\text{Bq}$$

It is named after the Nobel prize winner Madame Curie, who discovered many properties of radioactive elements. She also discovered an element which she called polonium after the country in which she was born.

Smoke detectors contain very small amounts of a radioactive isotope. When smoke enters the detector it is ionized by the radiation and this causes an electrical current to flow between two detection plates and the alarm goes off. The element often used is americium, an artificially made element.

Food and instruments can also be sterilized by radiation treatment with no overall harmful effects.

12.6.2 Rems and rads

Another unit that appears in some texts is the 'rad' which is 10^{-2} J of energy deposited per kilogram of tissue. Because different particles and radiations produce different damage, both the type and dose of radiation and effectiveness to produce biological damage must be considered. The unit 'rem' is sometimes used:

$$\text{rem} = \text{rad} \times \text{RBE}$$

where RBE is the relative effectiveness of the radiation to cause biological damage. The body mass of the person must also be considered.

A person living a 'normal' lifestyle would be receiving approximately less than 200 mrem per year. Usually only exposures above 25 rem (25 000 mrem) are considered to have any detectable effect upon human tissue.

The recognized units of dosage is the sievert. This unit measures the exposure to all forms of radiation, both radioactive and that from outer space, cosmic radiation. The average person in UK receives about 2.5 mSv radiation per year.

$1 \, \text{Sv} = 1 \, \text{J/kg},$ i.e. the amount of energy transferred to a specimen.

$1 \, \text{rem} = 0.01 \, \text{Sv}$

12.7   Conclusion

- All elements have isotopes. The vast majority of isotopes of elements are stable and not radioactive.

- The nuclei of some isotopes of some elements are unstable and give off radiation – they are said to be 'radioactive'. Isotopes that are radioactive are called radioisotopes.

- We are all surrounded by naturally occurring radiation, some coming from the rocks and some from space.

- We can use radiation to treat some of the effects of disease.

- A measure of the degree of radioactivity is the 'half-life'. Smaller values usually mean greater care must be taken with the materials.

- A radionuclide is an atom that undergoes radioactive decay.

- When isotopes disintegrate they give off radiation and change into a different element.

- The usual yearly dosage of radiation from any source for humans is 2.5 mSv.

Answers to the diagnostic test

1.

   i. The number of protons an element has   (1)

   ii. Both exist on the nucleus of an atom. Protons have a positive charge. Neutrons have no charge   (2)

   iii. Elements with the same atomic number but different numbers of neutrons   (1)

   iv. $^{4}_{2}He^{2+}$, nuclei of the helium atom shot out of the nucleus of a radioactive element   (1)

   v. $^{0}_{0}e^{-1}$, A fast moving electron, shot out of the nucleus of a radioactive element   (1)

   vi. Harmful radiation like X-rays shot out of the nucleus of a radioactive element   (1)

2. Summarize the properties listed in this chapter   (5)

3. This is the time it takes for a radioactive element to go from a certain value of activity to half that value. This time is characteristic of the element concerned and makes it of use when dating pieces of rock or ex-living material   (5)

4. Value in chemical analysis; tracing passage of medicine through the body; radiotherapy for treating cancers; PET (positron emission topography)   (3)

Further questions

1. Uranium 238 has a half life of 4.5×10^9 years and decays by the emission of an alpha particle followed by two beta particles. The new atom formed is also an isotope of uranium. Show these changes and what happens to the elements at each stage of decay.
   What can be said about the harmfulness of the radiation coming off at each stage? A further decay of five alpha particles ends up with an isotope of lead. What will be the atomic mass of this isotope of lead (atomic number 82)?

2. 'Radioactivity is a perfectly natural process and occurs all around us.' Explain this statement. Summarize, in table form, the properties of the three main types of radiation coming from radioactive sources.

3. (An open-ended question.) Investigate how radioactivity is used in your establishment in everyday use and laboratory use. Is there a Radiological Protection Officer in your organization, i.e. a person responsible for overseeing the use and preventing the abuse of radiation devices? Ask the people wearing the small photo badges why they wear them and how often they have to be tested.

4. Explain the difference between iodine 123 and 131 in both the nuclear arrangement of protons and neutrons and the difference in the break-down of the nucleus. The usual nonradioactive isotope of iodine is I 127. If 131 iodine gives off one beta particle what new element is formed?

$$^{131}_{53}I - \beta \text{ particle} \rightarrow ?$$

What will be its atomic number and mass number (mass of protons and neutrons in the nucleus)?

Reference

1. National Radiological Protection Board (NRPB). Tel: (01235) 831600.

13 Rates of Reaction

Learning objectives

- To stress the importance of rates of reaction for metabolic processes.

- To understand the factors that affect rates of reaction.

Diagnostic test

Try this short test. If you score more than 80 % you can use the chapter as a revision of your knowledge. If you score less than 80 % you probably need to work through the text and test yourself again at the end using the same test. If you still score less than 80 % then come back to the chapter after a few days and read again.

1. What effect does increasing the temperature have on the rate of a chemical reaction? (1)

2. Why is fresh food put in a refrigerator or frozen food in a deep freeze? (1)

3. Why in mortuaries are the bodies kept in refrigerated cabinets? (1)

4. Why does a piece of wood need a flame before it catches fire, then continue to burn? (1)

Chemistry: An Introduction for Medical and Health Sciences, A. Jones
© 2005 John Wiley & Sons, Ltd

5. What are free radicals?   (1)

6. Which would do your skin the most harm, a splash from concentrated sulfuric
    acid or dilute acid?   (1)

7. A chemical reaction was proceeding very slowly. Then the technician added a
    substance X to it and the reaction went faster. What is the general name of
    chemicals that behave like X?   (1)

8. What are the materials called that speed up chemical reactions in the cells of
    our bodies?   (1)

9. What do we call a chemical reaction that gives out heat?   (1)

10. What would happen to the speed of the chemical reaction in question 9 if it
    was heated?   (1)

Total 10 (80 % = 8)
Answers at the end of the chapter.

In 2004, two boys were playing on the ice of a frozen lake in Austria.
There was a sudden sharp crack and the boys disappeared into the icy water.
The police and ambulance service were called by some people nearby. The

emergency services arrived and located the bodies. By this time they had been underwater for almost an hour. After intensive artificial respiration for 25 min the boys' hearts began to beat again, weakly. A helicopter took them to hospital. The doctors very slowly raised their body temperatures and the boys eventually made a full recovery. The doctor explained that the cold considerably reduces the need for oxygen, therefore slowing their metabolism (and catabolism). This phenomenon is called 'mammalian diving reflex', which humans share with other animals, including whales, sharks and penguins. The throat closes and breathing stops, the heart rate is lowered and blood is circulated only to the body's essential organs (brain, lungs, heart). Only a small spark of life remains, which in cold water can last about 45–60 min. (Summary of a report in *Sunday Express* 21 March 2004, 13.)

13.1 Effect of temperature on reactions and metabolism

The degradation of the body following death is due to several factors, all of which depend upon chemical reactions. The body's metabolism is made up of two distinct types of processes: *catabolism* and *anabolism*.

Catabolism is the breaking down of molecules of foodstuffs into smaller molecules. This continues for a short time after death when, lacking foodstuffs and oxygen, the body catabolizes itself to some extent. It also continues during starvation – the bodies of famine victims essentially try to consume themselves. Long-distance runners eat a lot of pasta the day before a race to ensure they have enough carbohydrate energy to complete the race. If they do not, their metabolism runs out of instant energy and starts to catabolize the reserve carbohydrates and fats. This takes time, is not as efficient and is slower in energy conversion. Some diets work on this principle in order to reduce body fat.

Anabolism is the building up of the small molecules, once they have been absorbed, into bigger molecules which are needed for maintenance and growth of the body (see Chapters 4 and 5).

The boys mentioned at the start of this chapter survived because their bodies were very cold. The colder the conditions, the slower a chemical reaction takes place. In fact, a drop in temperature of 10 °C will roughly halve the rate of any chemical reaction. The boys' bodies were in water about 40 °C below normal body temperature. The first drop of 10 °C halved the rate of catabolism; the next drop of 10 °C halved it again. A drop in temperature of 40 °C gives a rate of catabolism that is a half of a half of a half of a half, i.e. one-sixteenth of the rate at normal body temperature. Hence, there was no detectable permanent brain damage. Fortunately it was not long enough for the water in the cells to begin to form ice crystals, which could have destroyed the cells and caused death.

Some lengthy surgical operations are performed with the patient wrapped in a cooling blanket to slow down metabolic processes, but not usually as low as the boys' temperature in this case.

13.2   Why does a chemical reaction slow down on cooling?

There are two reasons why a chemical reaction slows on cooling: first, for particles to react with each other, they have to collide. The lower the temperature, the slower the particles are moving. The slower they move, the less often they collide and the less energy the particles possess, so the slower the reaction goes. Second, for any chemical reaction to work, the old bonds in the reagent molecules have to be broken before new bonds can be formed to give the new products. Consider the simple chemical reaction

$$H_2 + O_2 \rightarrow H_2O \tag{13.1}$$

When hydrogen reacts with oxygen to form water, what happens is this:

Step 1 – the hydrogen and oxygen molecules collide;

Step 2 – the energy of the collision causes the bonds of some of the hydrogen and oxygen molecules to break, forming hydrogen atoms. We call these free radicals.

$$H{-}H \rightarrow 2H^* \tag{13.2}$$

The asterisk shows that this is an atom and has free un-bonded electrons. Here one hydrogen *molecule* splits up to give two hydrogen *atoms*.

$$O{-}O \rightarrow 2O^* \tag{13.3}$$

(One oxygen molecule splits up to give two oxygen atoms containing unbonded electrons).

Step 3 – the hydrogen and oxygen atoms form new bonds to produce water molecules.

$$H^* + H^* + O^* \rightarrow H_2O \tag{13.4}$$

These atoms with their spare electrons due to breaking open the covalent bond are termed 'free radicals'. They are very short lived as further collisions occur either to reform the original molecules or to join up to form the new compound.

Here are more details of the steps. If the original molecules do not split up, no new molecules can be formed. Now, any real sample of a reagent will contain billions of molecules. At any given temperature, not all of these molecules will be moving at the same speed. The temperature is a measure of the *average* speed they are moving at. So some collisions will be between fast-moving molecules and some between slow-moving ones. Some collisions will not be energetic enough to break the bonds in the molecules, so no reaction will occur. In other words, not all collisions result in a reaction. A reaction will only occur if the energy of the collision is greater than a certain amount. This amount will vary according to the strength of the bonds in the reagent molecules. The name given to the energy needed to break the old bonds and initiate a reaction is 'activation energy'. Because different molecules have different strengths and numbers of bonds, the activation energy will be different for different reactions.

At lower temperatures, molecules move more slowly so there will be fewer collisions in which the average energy exceeds the activation energy and the reaction will go more slowly and less product will be formed. At higher temperatures there will be more molecules with the energy to overcome the activation barrier. The particles have more overall energy to transfer at higher temperatures.

A piece of wood will only burn in air if it is given a start with a flame so that some molecules are given extra energy, and more collisions produce broken bonds (or exceed the activation energy). The heat of the flame, once started, continues to give out heat as the reaction is exothermic. This supplies the rest of the energy for the wood to burn.

To start a chemical reaction, the initial energy barrier must be overcome to reach the top of the energy curve at point A in Figure 13.1. At this point the free radicals or hydrogen and oxygen atoms are formed. After this, some of the atoms react to form the product of water and energy is given out. This is shown by the downhill path of the latter part of the graph. It can be seen that more energy is given out than is needed to initially start the reaction. This means that heat is given out in this reaction. Some of this heat breaks open more molecules and the reaction carries on until all the materials have been used up.

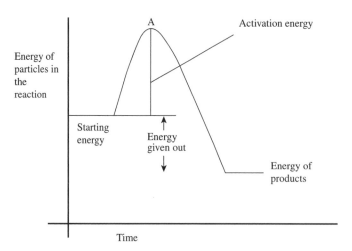

Figure 13.1 A chemial reaction showing an activation energy barrier

13.2.1 Implications of the graph in Figure 13.1

- Following death, the decay processes (all of which depend on chemical reactions) proceed faster in hot climates than in cold, hence the need to bury or cremate bodies soon after death. This helps reduce spread of disease.

- We keep our food in fridges and freezers to slow down the chemical reactions that lead to decay. The molecules are deprived of the energy they need to overcome the activation energy to start decaying. If exothermic reactions start, then some heat is given out. This heat is counteracted by the low temperature in the fridge.

Most chemical reactions give out heat when they react and anything reacting with oxygen usually gives out a lot of heat. Reactions that give out heat are called *exothermic* reactions. Oxidation of glucose during respiration gives us energy and keeps our body warm.

$$C_6H_{12}O_6 + 6O_2 \rightarrow 6CO_2 + 6H_2O + \text{heat given out}(-\Delta H) \qquad (13.5)$$

The overall amount of energy given out during the reaction is shown by the difference between the initial energy of the reactants and the final energy of the products.

Endothermic reactions require heat to keep the reaction going. In this case the finishing energy value of the products will be above the value of the starting energy.

When the initial and final products have approximately the same energy then it is possible that the reaction will proceed in both directions with equal ease and so the reaction is easily reversed.

13.3 Free radicals

Free radicals are usually small groups or single atoms that have spare un-bonded electrons which are readily available for a chemical reaction. They are very short-lived items lasting less than a thousandth of a second but long enough to do harm to cells or produce a chemical reaction.

In our bodies oxygen atoms, O^*, and small peroxide radicals are made as a by-product of the respiration of carbohydrates, and these can do a lot of damage to living cells. The blood of our bodies is ready for this and contains enzymes that quickly deal with free radicals by reacting with them (see Section 6.3).

The presence or absence of these free radicals can change the 'cell clock' and affect the rate at which cells divide. The production of unwanted free radicals inside some cells makes them divide at a rapid rate. In these circumstances some cells make abnormal or cancerous growths. If there are too many free radicals the cells are killed altogether.

Free radicals are made when a normal covalent bond splits in half with each atom receiving an equal share of the electron pairs. For example a hydrogen molecule H—H splits into two equal halves with each hydrogen atom receiving one electron, H^*. Covalent bonds that split in half in this way are said to break 'homolytically'. Free radicals are often shown by having an asterisk on their formulae (see also Section 13.2).

13.4 Effect of concentration on chemical reactions

In 1997 there was an urgent phone call from a chemical works to the ambulance service. There had been an accident in the laboratory at the works. A flask had broken and the solution of phenol it contained had sprayed out onto the forearm of the worker. It was not much, but it was enough. The worker was, of course, not wearing goggles, nor a laboratory coat. Come on, it was summer – it was hot! Tee-shirts were the order of the day. Anyway, 'it will never happen to me'.

By the time the ambulance arrived at the works, blisters 12 cm across had formed. Fortunately the outcome was not fatal, although when she goes to discos nowadays, that particular worker tends to wear long sleeves to hide the scar tissue. She also wears every possible piece of protective clothing for laboratory procedures. Because it *can* happen to me.

Roughly a hundred years earlier Sir Joseph Lister suffered exposure to exactly the same total amount of phenol with no blisters and no scarring at all. Sir Joseph conducted surgical operations under a fine spray of a very dilute solution of phenol. The speed of a reaction in solution depends on the concentration of the solution. Thankfully there are better ways of ensuring hygiene in operating theatres now as phenol is nasty stuff.

The two people were exposed to the same chemical, phenol, but one was more concentrated than the other. Concentration makes a difference to the speed and extent of a chemical reaction. The chemical worker had been hit by a very concentrated solution of phenol and Lister by a fine spray of dilute phenol solution.

13.4.1   The effect of concentration on the rate of reaction

Let us define *concentration*. The more solute you dissolve in a solvent, the more concentrated the solution is. When no more solute can dissolve at a particular temperature we say the solution is saturated. More precisely concentration could be expressed as the number of grams of solute in $1 \, dm^3$ (or litre) of solvent. More usually it is expressed as the number of moles (molecular mass of the substance in grams) of the solute dissolved in $1 \, dm^3$ (or litre) of solution. Remember, when one molecular mass of a substance is dissolved in $1 \, dm^3$ we say the solution is 1 M.

The more concentrated a solution is, the more of the active chemicals are present ready to react and so the faster the reaction. For example, the more concentrated a solution of orange juice, the more it reacts with the taste buds. A neat gin is more concentrated than a gin and tonic. Two spoonfuls of sugar in tea make it more concentrated than one spoonful.

13.5   Catalysts and enzymes

Catalysts and enzymes in very small amounts greatly alter the speed of a reaction. They cannot make impossible reactions work, but can alter the rate of those reaction that are possible. They do not alter the product of the reaction, only the speed you make them. Providing the catalysts are not destroyed or poisoned they can be used over and over again. How can such tiny amounts of a substance be so important? As you know, any chemical reaction happens in several steps. We hesitate to repeat them, but we will:

Step 1 – the molecules collide.

Step 2 – if the energy of collision is greater than the activation energy, the old bonds break.

Step 3 – new bonds are formed, and new compounds are produced.

When a catalyst or enzyme is present, it works because it lowers the activation energy of the reaction, so less energy is needed for the particles to react together. Therefore the products are made more quickly (Figure 13.2). How does it do this?

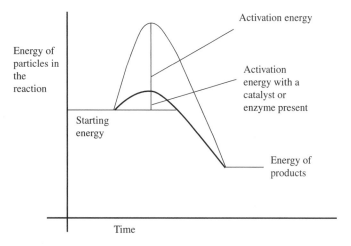

Figure 13.2   The effect of a catalyst on activation energy of a chemical reaction

13.6   How catalysts and enzymes work

There have been many theories of the exact mechanism of the action of an enzyme because so little of it is needed to affect the rate of a reaction. Some enzymes can speed up a reaction as much as 20 000 times. Some negative catalysts (or inhibitors) slow down the rate of a reaction.

One explanation for the mechanism is as follows: in a reaction the two reactants need to collide with each other many times for a reaction to occur. In doing this only a few of the molecules are in the correct orientation to react with their counter-part. Owing to the randomness of motion, some of the orientations will be suitable and the reaction can take place. However, in many thousands of collisions the orientation is not correct and no reaction takes place. When an enzyme or catalyst is present, one reactant fits onto the enzyme molecule in the correct specific

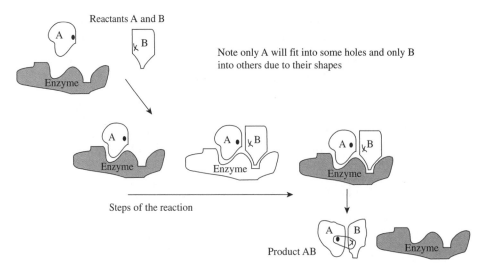

Figure 13.3   The mechanism of action of enzymes

orientation (Figure 13.3). This means that, when the other reactant comes close to the enzyme/reactant combination, it is immediately in the correct orientation and alignment for the two reactants to join up. These react immediately and then come away from the enzyme because the new molecules that are formed do not stick to the catalyst. The enzyme is available for reuse. Because the reactants are lined up correctly, no time is lost in useless random collisions between the reacting molecules. The mechanism is sometimes referred to as the 'lock and key' model.

Usually catalysts and enzymes are very specific and will only speed up one particular reaction. For example, one enzyme in the blood will only catalyse one type of reaction and will not affect other reactions. Therefore a 'blood clotting' enzyme reacting at the site of a cut will not be affected by a 'clot dissolving' enzyme at a different site. The reactions are very specific.

Our bodies contain a collection of sites where enzymes are secreted, each for a specific reaction. Each reaction only requires minute quantities of the specific enzyme for it to start reacting (see also Chapter 6).

When the body requires blood to clot, as in the case of a cut or wound, enzymes are involved. Thrombin gathers in the area of the cut or wound and catalyses the soluble protein in the blood to form the insoluble fibrin structure, thus helping clotting and wound repair. The thrombin speeds up the rate of the reactions involved in the formation of insoluble fibrin from soluble protein. This is a simple example, but it goes, we hope, to show how very tiny amounts of substances can be essential for the efficient working of all living systems.

13.7 Application of chemical reactions to drug use

The effectiveness of a drug or medicine depends on it getting to the site of a reaction as quickly as possible. It must not be broken down before it reaches the site. It then reacts with the chemicals at the site and hopefully the drug lasts as long as is necessary. The concentration of intake is often critical as not all the drug reaches the site at the same concentration as it was administered. Some will be lost to other tissues of the body and some might be broken down and some eliminated. The effectiveness of a drug depends on other factors, including the weight and age of the person, pH, temperature, water solubility, toxicity and removal of waste products.

The intake of a drug must be at a sufficient concentration that it has a therapeutic effect at the site needed, but not so high as to cause toxicity. Too little can be ineffective. The above factors all interact into a study called 'pharmacokinetics', which concentrates mainly on the four major factors:

- intake and absorption;

- transport and distribution;

- metabolic break-up;

- excretion.

These four factors are quantitatively studied to find the best way to administer the drug. Any drug administration should therefore be tailor-made for a particular person and the severity of the situation. For common colds and the like this is not usually required but for the more invasive cancer drugs this is crucial. Overdoses can have serious consequences. Similarly, miscalculation of radiation treatments can be catastrophic for the patient. The analysis by the pathology laboratories of periodic samples of excretion products can give a good indication of the metabolic break-up of the drug and its retention at the crucial sites.

Chemists studying the pathways, retention and metabolism of a drug often use animals to test the models for their modes of action before trying it out on humans. There is always a safety factor built into any human trial model as drug mechanisms and the human body are such complex systems. The models are expressed in mathematical equations and graphs that allow each drug and person to be studied accurately. Increasingly, animal tests are being replaced by computer models that simulate the conditions within human cells. That is why medical records, dosage rates and temperatures are all recorded accurately for each patient.

13.7.1   Intravascular administration of drugs

This method introduces the drug by injection or infusion and bypasses the 'absorption' factor as the drug is often placed directly at the site needed. This method can accurately maintain the correct concentration of the drug in the plasma, blood or tissues.

13.7.2   Extravascular administration of drugs, orally

Orally administered drugs are absorbed through the GI (gastrointestinal) tract membranes and small intestine. In these cases the drug is often most effective after a meal, as the presence of food slows down the passage of the drug and so gives it a better chance of being absorbed. Some of the drug is lost as it passes through the stomach and liver before it can reach the main distribution system or systemic circulation. Some of the drug's effectiveness might be lost at these stages due to metabolism in the liver. The amount of the drug that gets past the liver and enters the general circulatory system is called the 'bioavailability' of the drug. This is the fraction of the dosage that enters the circulatory system.

Any drug that does not pass through the liver with a high enough concentration of bioavailability will be of no use. Too high concentrations of some drugs can cause liver damage. This is often trialed in animals when the drugs are being tested for effectiveness. For some drug treatments it is necessary to know not only the bioavailability but also the retention time within the body so that a further dose can be administered to keep the effectiveness of the drug at a maximum, e.g. 'take twice a day after meals' instructions.

All these factors and many others are used in drug trials and development during and before clinical trials with humans. Legislation demands that accurate values are assigned to dosage, absorption, elimination, distribution, clearance, bioavailability and half-life (the time it takes for a drug to go from any concentration to half its value *in situ*). All these factors must be understood when administering a drug, and particularly a new drug, to a patient. The details are included in the literature of the drug both in the packets and in accompanying booklets of detailed instructions. The difficulty comes when the patient is taking more than one drug for more than one major complaint. That is when professional judgment and an understanding of the nature of the drugs being used is needed.

Answers to the diagnostic test

1. Increases the reaction rate (1)

2. To slow down decomposition reactions (1)

3. To slow down decomposition reactions (1)

4. It needs initial heat to overcome activation energy to start the reaction (1)

5. Small atoms of molecular groups with free electrons (1)

6. Concentrated (1)

7. Catalyst or enzyme (1)

8. Enzymes (1)

9. Exothermic (1)

10. It would slow down (1)

Further questions

1. i. What exactly does the following equation tell you and what does it not show?

$$A + B = C + D$$

ii. What would be the effect of applying external heat to an exothermic reaction? Would it help it or hinder it? Why are cold packs applied to a person with a fever? Why are some operations conducted while the patient is surrounded by a cooling jacket?

2. Explain how the idea of activation energy applies to the simple process of striking a match.

3. A sugar cube will not burn on its own, but if some cigarette ash is sprinkled over it, it will burn. Explain this process in terms of catalysis.

4. Many body reactions work more effectively in the presence of an enzyme. Give examples and explain how an enzyme works.

5. Explain the following sets of observations in terms of chemical reactions.

   i.   Vinegar is put on chips but not sulfuric acid
   ii.   Absolute alcohol is used to store specimens in the patholog laboratory but alcoholic drinks are sold in the bar.
   iii. Car exhausts have catalytic converters to reduce pollution.
   iv.   Refrigerators and deep freezers are used to store specimens for a longer time, including human embryos, which are stored in liquid nitrogen.
   v.   Equipment in hospitals is sterilized by high temperature and pressure treatment.
   vi.   Equipment and hands must be washed thoroughly in disinfectant.
   vii. Allowing nurses to wear the same clothes to go home through crowded streets as they wear around the hospital is a hazard to public safety and hygiene.
   viii. The warm temperatures of a hospital ward makes ideal conditions for bacteria to grow. Any moist areas, like around taps and sinks, are particularly suitable for bacterial growth.

6. What would be your advice to a hospital manager wanting to eliminate MRSA from the hospital? How much of this advice depends upon the use of chemicals?

14 Overview of Chemicals Fighting Diseases

Learning objectives

- To give an overview of historical and present day information regarding topical areas of medical chemistry.

- To supply a source of information and stimulus for further reading and research in relevant journals.

There are no diagnostic questions for this unit as most of the material will be new to the reader.

14.1   Drugs ancient and modern

The same sources of medicine are available to us today as were available, even to ancient civilizations. They is the wide array of chemicals found in plants. Witch doctors, medicine men and old wives proved to be very perceptive in knowing which plants to select for which complaint.

What is the potent poison speared on arrow heads used by South American natives when hunting? Analysis of the materials shows it is curare. This has muscle relaxant and neuromuscular blocking effects. These properties have been used in modern medicine and surgery.

Chemistry: An Introduction for Medical and Health Sciences, A. Jones
© 2005 John Wiley & Sons, Ltd

In 1785 William Withering showed that an extract from foxgloves (*Digitalis purpurea*) was beneficial for treating people with heart problems. The majority of cures for disease in our present society have come from nature either by extracting the active components from the plants or by mimicking them using chemical synthesis. Both of these are very fruitful areas of research.[1]

Malaria is a worldwide problem. Even in Britain there are some marshes where the malaria mosquito exists. The extract from the cinchona tree was the only known treatment as early as the seventeenth century. The action of the ingredient, quinine, was not fully understood until about 1820. A synthetic drug was made in 1930 to mimic quinine in its action and so avoid the tedious job of collecting and extracting the natural quinine (Figure 14.1).

Figure 14.1   Quinine

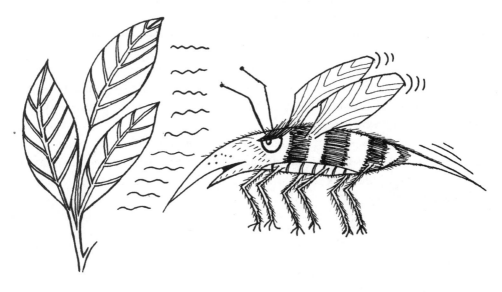

Keep off that cinchona tree, its smell is killing me!

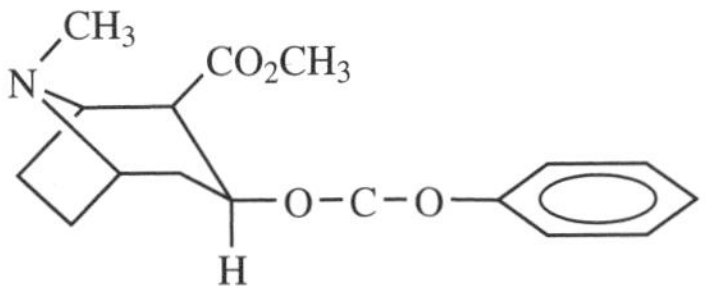

Figure 14.2 Cocaine

In 1884 cocaine (Figure 14.2) was introduced to Europe by Carl Koller as a local anaesthetic for minor operations. The drug is more in the news today for its abuse by drug takers.

Alfred Nobel, the explosives expert, invented the process for making nitroglyerine, which is a powerful explosive. In small quantities it was shown to help angina patients. It was introduced as a treatment in 1878 in Westminster Hospital. Nobel would not try the treatment himself and died of a heart attack. Better heart drugs are now available.

Queen Victoria had an abscess removed in 1870 under an anaesthetic of ether (diethyl ether, $C_2H_5 \cdot O \cdot C_2H_5$). She also used the pain killer morphine (plant extract) and an antiseptic containing phenol. In 1861 her husband died of typhoid fever at the age of 42. Today an antibiotic would have easily cleared up this disease. In the nineteenth century a large number of deaths occurred due to bacterial or viral infections.

There was a worldwide influenza epidemic in 1918–1919 which killed about 20–30 million people throughout the world. There was no known cure for this disease. Is there a cure at the present time?

One of the early pain-relieving drugs, aspirin, was introduced in 1899 by Felix Hoffman, working for the German pharmaceutical firm Bayer. In 1898 Bayer also introduced heroin as a pain killer. Aspirin[2–4] is a man-made synthetic chemical compound, acetylsalicylic acid. It is made from salicylic acid. The ancient Greeks and native South Americans, among others, found that the bark of the willow tree eased fever and pain. We now know that this is because the bark contains salicylic acid. Salicylic acid is bitter and irritates the stomach (see also Chapter 1).

Aspirin is now the most widely used drug for fever, mild to moderate pain, and inflammation due to arthritis or injury. It is an effective analgesic. Aspirin causes a very small amount of gastrointestinal bleeding that can, over time, cause a slight iron deficiency. In some people it causes gastric ulcers if used long-term. Complications can be avoided by using enteric-coated aspirin, which does not dissolve until reaching the intestine.

Aspirin must not be given to children who have chicken pox or influenza because it increases the risk of them contracting the rare and frequently fatal Reye's syndrome, a disease of the brain and some abdominal organs.

Aspirin acts by interfering with the synthesis of prostaglandins, which are connected to the causes of inflammation and fever. In particular it is an inhibitor of the cyclo-oxygenase (Cox-2) enzyme. Studies of the use of aspirin as an anti-clotting agent suggest that half an aspirin tablet per day may reduce the risk of heart attack and stroke in some people.[3]

- Aspirin is probably the wonder drug of the twentieth century. More aspirin is used than any other drug (an estimated 50 000 million tablets per year are used worldwide). It is good for pain relief, and thinning the blood after heart attacks. Taken in small quantities it helps to prevent heart attacks.

- It has recently been reported that aspirin can also help in the prevention of prostate cancer in males. Research on this possible use is continuing.

- In addition, 125 mg of the herbal alternative to aspirin, namely pine bark extracts[4] containing bio-pycnogenol, have the same effect as 500 mg aspirin as an anti-blood clotting agent. It is an alternative for people who have stomach ulcers.

- The search is still on to find a drug as effective and far-reaching as aspirin but which does not produce the side effect of slight stomach bleeding. Perhaps the Cox-2 group of compounds might be its successor.

At the beginning of the twentieth century chemistry was becoming a growth industry and many could see the value of plant extracts for curing diseases. These were analysed and their active ingredients were synthesized. In 1923 a plant extract from a plant used in Chinese medicine led to the synthesis of the anti-asthma agent (beta antagonist) salbutamol (Figure 14.3).

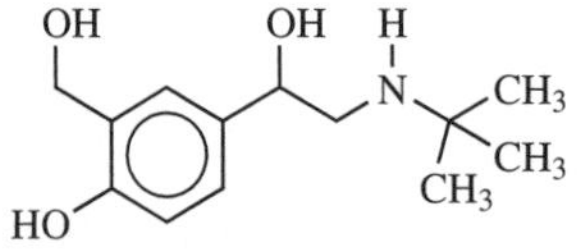

Figure 14.3   Salbutamol

In the early part of the twentieth century the synthesis of sulfonamide drugs such as anti-bacterials led to the first effective treatments of the fatal diseases septicaemia, bacterial meningitis and the killer disease, pneumonia. The active form of the drug is *p*-aminobenzenesulphonamide (Figure 14.4).

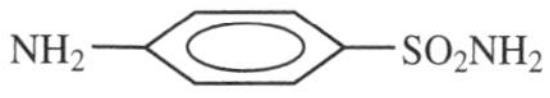

NH$_2$ —⟨◯⟩— SO$_2$NH$_2$

Figure 14.4 *p*-Aminobenzenesulfonamide

Serendipity occurs all the time in science. People like Fleming must be applauded. The famous discovery of penicillin by Fleming in 1928 is a well-known story. He noticed the effect of a mould killing bacteria. The derivatives of penicillin must be some of the greatest discoveries ever made in medical science. We now know that there are a series of penicillin derivatives based upon the structure shown in Figure 14.5.

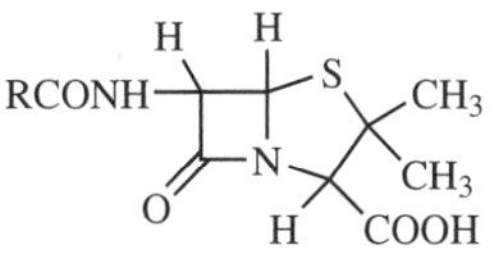

Figure 14.5 Penicillin

Unfortunately the bacteria have also adapted to become resistant to various types of penicillin. New areas of research and plant searches are being undertaken to find a completely new way of attacking bacteria, particularly the very resistant ones of methicillin-resistant *Staphyloccus aureus* (MRSA). Recently a newspaper[5] reported a new line of research: 'scientists have managed to extract, from rock pool slime, a chemical that stops dead MRSA.' Anti-cancer and anti-inflammatory drugs are being sought from the same source.

Many natural products have been investigated to try to kill MRSA. These include honey.[6] It is probably the small amount of hydrogen peroxide present in honey that is the active component. Garlic was tried with 200 patients and found to be effective in killing off MRSA in the majority of cases.[7] All natural remedies have to be thoroughly investigated to confirm that they do not interact with any drug the patient may be taking.

14.1.1 Other natural sources

So far we have described the use of nature for curing diseases. This still continues as probably only 10 % of the world's plants have been screened for their possible chemical and curative properties. A different approach was adopted by James Black at ICI. He decided to consider making a drug to order: a designer drug. He and his team set out to design a drug that would selectively block the beta adrenergic receptors on the heart muscle. This would reduce heart pain for people with angina or heart-related problems caused by the stimulation of the heart by adrenaline released when under emotional or physical stress. This led to the manufacture of beta blockers.

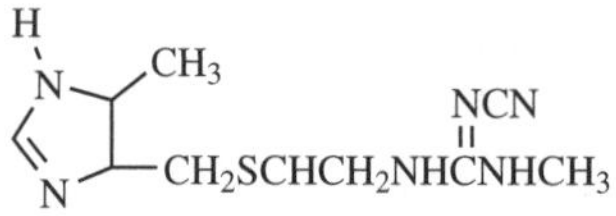

Figure 14.6   Cimetidine

Cimetidine, sold as Tagamet (Figure 14.6), was also another drug specifically designed to block the release of histamine from stomach cells. This drug is a good remedy for stomach ulcers. Cimetidine, discovered in 1970, was the first effective anti-ulcer drug. It works by lowering acid secretion in the stomach, so helping healing. About 1000 tonnes of the drug are manufactured each year.

In the 1930s research on the isolation of progesterone, a steroid and hormone, from natural sources led to synthetic pathways for their manufacture. Research led to the development of steroids used as contraceptives. Others were developed as anti-inflammatory steroids. Other steroids have been developed to treat breast cancer and another for prostate cancer.

14.2   Cancer treatments

Cancer chemotherapy has led to the development of taxol, a widely used anti-cancer drug. Taxol is very poisonous to tumours but it also affects almost everything else that grows in the body.

Cancer is caused when a cell is genetically damaged and starts to divide too quickly. Taxol slows down and even prevents the cancer cells from dividing to form new cancer cells. Originally taxol was made from the bark of the Pacific yew tree but it took almost two trees to produce enough pure taxol to treat one patient. Taxol is now synthesized by a multi-step process but it is still expensive.[8]

When a tumour is detected there are a number of treatments possible, ranging from its surgical removal to the use of specific chemicals to target the tumour. One synthesized platinum compound, cisplatin (Figure 14.7), is widely used for treating

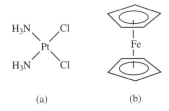

Figure 14.7 (a) Cisplatin and (b) ferrocene

testicular, ovarian, bladder and neck cancers. It is administered intravenously. It diffuses from the blood into the cytoplasm of cells. Here it forms compounds with the nitrogen on guanine groups on adjacent strands of the DNA of cells. This distorts its shape and structure. Unfortunately cisplatin is not selective for just cancer cells. It can lead to severe side effects, including renal impairment and loss of balance and hearing which are often only partially reversed after treatment.

The search for other, less toxic, water-soluble specific drugs has led to a group of compounds known as 'sandwich compounds' being investigated, e.g. ferrocene. These compounds have a metal atom held between or alongside suitable plate-like molecules which under suitable conditions can be attached to DNA chains and distort their structure. There is evidence that some of these compounds are better than cisplatin for killing off some cancers but less effective for others. It is possible that a method of combining suitable drugs in a 'cocktail' could be effective so that what one drug misses the other one finds. The mixtures must be carefully formulated and possibly made up individually for each patient. This is called 'combination therapy'.[9]

One of the most widely used treatments for breast cancer uses tamoxifen (Figure 14.8). It was originally developed as a contraceptive, but it was found that it inhibited oestrogen uptake by cells. Tamoxifen alters the shape of a protein on the surface of the oestrogen receptor and so stops it from giving the signal to 'grow'.

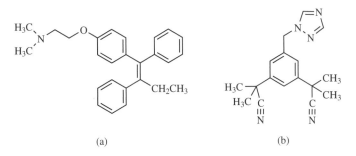

Figure 14.8 (a) Tamoxifen and (b) anastrozole

There are side effects from taking tamoxifen, including increased risk of blood clotting and damage to the lining of the womb. A replacement drug has been developed called anastrozole, which acts by reducing the amount of oestrogen produced. The only side effect is an increased risk of osteoporosis.

Natural 'stinkweed' extracts could replace the more toxic anti-cancer drugs. Japanese research showed that extracts from it, when injected into malignant brain tumour cells grown in the laboratory, brought their growth to a standstill and they lost their ability to spread.[10–12]

Chemicals when taken in tablet form can also attack areas of the digestive system. Work is in progress to prevent this happening. One method is to develop a harmless virus which would act as a 'Trojan horse' to smuggle genes to the exact spot needed. It would then release its contents on an instruction from a chemical only present at that site.

An anti-cancer drug based on cisplatin carries its active ingredients to the site required and this is then illuminated to activate the drug. Any drug compound that is carried to other sites is not illuminated and so its side effects on healthy cells are eliminated. Perhaps less hair loss, sickness and nausea will be experienced as a result.[11] Some research teams think they can 'knock out' the cancer cells with specific enzymes,[13] whereas other teams are focussing on the energetics of the cancer cells and trying to switch off their energy sources.[14]

The rapid advances in analytical chemistry and the detection of ever smaller quantities of material involved in the processes in our bodies mean that our understanding of the metabolic mechanisms and interactions with drugs is becoming more and more sophisticated. The small quantities of carbon monoxide and its presence in the brain, the massive role of nitric oxide and the detection of minute quantities of oxidants and anti-oxidants in the blood and cells are discussed in other chapters.

Research and development of drugs for Alzheimer's disease,[15] heart disease, AIDS, arthritis, eating disorders, psychological disorders, drug abuse, skin cancers and asthma are all in progress. Each month reports of new discoveries are made and, by the time you read this, many more new avenues of research will have opened up. Extract of lemon balm is being tested as a possible material to counteract the symptoms of Alzheimer's disease.[16] The next discovery might be the use of garlic as an anti-bacterial agent, anti-oxidant or for lowering cholesterol, Ginseng for treating fatigue or as an immune system booster, friendly bacteria to aid diabetics, pectin in citrus fruits as a possible source of anti-cancer properties, aricept or exelon for treating Alzheimer's, St John's wort for treating depression as a replacement for Prozac, clover extracts for cure of HRT, cannabis to help people with MS or, inhaling small quantities of a gas containing nitric oxide to supplement our systems – after all it is used for premature babies!

There has been progress in the development of anaesthetics from the primitive use of excess alcohol, through to using ether or nitrous oxide gas (N_2O, laughing gas) and in the 1950s the development of the halothanes (these had the side effect of causing liver hepatitis in some patients), leading to the more recent use of fluoro compounds including enfluorane and isofluorane. The nicer smelling sevofluorane is now used. These give faster, safer induction and easier recovery. A future anaesthetic will probably be a continuous injection of a suitable analgesic agent and hypnotic, controlled by computer connected up to the patient's blood and breathing systems.[17]

14.3 Pain killers

Arthritis or osteoarthritis is suffered by over 1 million people in the UK alone. These are joint diseases that affect the cartilage of 85 % of over-70-year-olds. Paracetamol and NSAID (non-steroidal anti-inflammatory drugs) can be used to relieve the pain, but it is not a cure. Other treatments include the use of cortisosteroids and glucosamine (Figure 14.9). The NSAID drugs cause ulcers and bleeding. Drugs being introduced to help alleviate these side affects include celecoxib, rofecoxib and vioxx. These drugs have come under review in 2005 due to some unforeseen side effects and have been removed from use.

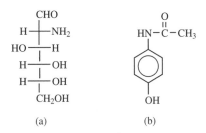

Figure 14.9   (a) D-Glucosamine and (b) paracetamol

Rheumatoid arthritis affects 600 000 people in the UK. It causes inflammation and thickening of the membrane lining the joints. Its cause appears to be genetic or inherited. The treatment is the same as arthritis and also the use of the slightly toxic DMARDs (disease-modifying anti-rheumatic drugs). Newer drugs have been developed for blocking or inhibiting the body's Cox-2 (cyclo-oxygenase) compounds. These produce the prostaglandins made by the inflammatory reaction.

The race does not stop, since now the challenge is to eradicate the worldwide killer virus and related effects of AIDS. AIDS progressively destroys the human immune system. The treatment being used is a mixture of drugs to combat the action of the virus. Even if a cure is found, would the poorer African nations, where this is rife, be able to afford it?

Perhaps one day someone will devise a safe alternative to smoking and so save the lives of smokers and secondary smokers. Maybe one day an effective treatment will also be found for the common cold.

14.4 Stopping attacks by 'aliens' on our bodies: viruses and bacteria

Our survival is dependant upon our bodies being able to detect attack by viruses, bacteria and dangerous chemicals. Our bodies contain a mechanism to detect and attack any foreign materials. This is our immune system. It consists of white cells in the blood of two main types. These are the B- and T-type detectors. Other cells that are present in the blood are the phagocytes, which consume intruders, the eosinophilis, which make poisoning chemicals to put off intruders, and mast cell and basophils, which make inflammation-causing chemicals. The foreign intruders meet an army of defending forces. All these actions are initiated by chemical reactions in the body proteins.

The surface of a bacterium, for instance, is covered by proteins in twisted and dented shapes with chemical groups sticking out all over it. All bacterial antigens have their own characteristic shape. The B-cells of our bodies recognize these intruder proteins as having opposite shapes and charges to themselves and therefore can bind or react with them. Each B-cell has one single Y-shaped part matched to one Y-shaped part of the antigen. The B-cells manage to overpower the antigen by sheer numbers. Each B-cell can recognize about 10^8 different shapes.

Once the antigen has been attacked by the B-cells it releases antibody molecules called immunoglobins (or Igs). These have the same Y-shaped protein and are exact copies of the receptor protein, which means they can join up to any antigen with the matched shape. In doing this they leave the tail part of the Y sticking out, which is recognized by the phagocytes, which eat the organism. This whole attack is managed and controlled by two types of T-cells, the helper cells which help in antibody production, and T-supressor cells, which slow down antibody production.

It is thought that the T-cell balance can be different in different people. If T-helper cells predominate, then their immune system can be hyperactive and react even with relatively harmless particles like dust or pollen. This causes the mast cells to release chemicals including histamine and other chemicals that cause inflammation, dilation of blood vessels and the production of mucus. This occurs in the nose for hay-fever sufferers. Reaction in the air ducts causes an allergy to, say, dust or dust mites. Release of these chemicals can cause eczema or asthma.

For some people allergic reactions can be caused by certain foods. The mast cells can be triggered at any point in the journey of the food through the body. Some areas of the body are more sensitive in some people. Some people can be sensitive to wheat (or the gluten in wheat, causing coeliac disease), eggs, soya, peanuts or shellfish. The most dangerous reaction to food is anaphylactic shock. This is when the antigens in the food enter the blood and cause a sudden drop in blood pressure, narrowing of airways and swelling of the throat. These reactions can be rapid and fatal. Food allergy is not the same as food intolerance, which is mainly due to the digestive system.

The treatments for allergies try to lessen their effects by using anti-histamine drugs or, for asthma, using salbutamol, which relaxes muscles in the airways. Research is still continuing to develop new drugs to help allergy sufferers, whose numbers seem to be increasing.[18]

14.4.1 Anti-viral treatments

Anti-viral drugs use mechanisms that interfere with nucleic acid and protein synthesis, inhibiting their attachment to and penetration of host cells. Because the viruses frequently adapt their structures, their elimination is difficult. A research team in Cambridge[19] are looking at what they call a 'mutator protein'. This protein is made by our own cells and is released to 'sneak' inside certain viruses and cause chaos and mutations in their genome. Unfortunately the AIDS virus has evolved a defence against this process, but it is useful for other viruses.

Another virus, of the coronavirus type, has made headline news under the name SARS. It seems to be a disease that has crossed species boundaries, from animals to humans. This disease appears not to kill its victims directly but sets up a storm of immunochemicals in the body called cytokines. These are inflammatory materials and can be more deadly than the virus itself in large quantities. Influenza and pneumonia use the same technique.[20] The 'bird flu' of 1997 and 2004 seems to have crossed the animal–human species barrier and has created an urgent need for research into counteracting this problem.

14.5 AIDS and HIV

The treatment of AIDS (Acquired Immune Deficiency Syndrome) is one of the major areas of research at present. In 2002 it was estimated that there were approximately 14 million reported cases of AIDS and probably many thousands not reported. It was first noted as a disease in 1969 when analysing blood samples. In 1981 New York reported its first major outbreak and in 1982 the Centre for Disease Control first coined the acronym AIDS.

Its mode of action seems to be along these lines. When the thin skin layers and mucus of the mouth, anus or vagina are damaged then, as in all cell damage, there is a large influx of white cells to combat any possible invaders. It is precisely this gathering of white cells, normally effective for killing off viruses or bacteria, that HIV (Human Immuno-deficiency Virus) attacks and takes over.

HIV attaches itself to the surface of the white cells and then empties itself into the white cells and hijacks them to produce more HIV. Thus the more white cells that arrive to attack the 'intruder', the more it likes it and more quickly it reproduces. It has a high rate of reproduction.

It sometimes takes as long as 6 months before antibodies to HIV can be detected, and a person is said to have AIDS when the normal immune system has been overcome. This makes the person vulnerable to the attack of other diseases. AIDS victims therefore usually die from these secondary diseases, like pneumonia or tumours, rather than from the action of HIV.

The early drugs that were used merely lengthened the life-span of patients. The virus seems to be very clever in adapting to any single new drug used. A three-drug mixture was used to confuse the virus and prevent it from adapting or mutating to the change of drug, but these can lead to nasty side effects, including loss of feeling in limbs, diarrhoea and nausea. Research is continuing to develop a vaccine against HIV because the virus is a genius at mutating. A different vaccine might be needed for different countries. One vaccine that is being trialled stimulates the T-cells to attack the HIV.

The scale of the problem is enormous and it is estimated that there are 40 000 new cases of AIDS in the USA each year. Some of the African countries have a large percentage of their population with HIV in their system or full-blown AIDS. The process of aggressive education in communities from primary schools to mature adults is proving its worth. The emphasis of one partner, marriage and sexual responsibility is helping to reduce the incidence, but it will take several generations to stamp out this plague as children are still being born with HIV in their blood. In her presidential address to the Association for Science Education in 2003[21] Gill Samuels reported that 'the spread of the drug-resistant virus is inexorable and of significant concern. It is calculated that in 2005, 50% of new HIV infections in San Francisco will be with the drug-resistant virus. Thus investment in drug development and disease control is vital'. Research and development is progressing so quickly that up-to-date medical journals need to be scanned regularly.

14.6   Gene therapy

Gene therapy is a valuable tool for manipulating or modifying genes to combat many genetic diseases. Huntington's disease, a fatal brain disease, is thought to be combated by a method using RNA interference. This method uses short interfering pieces of double-stranded RNA (called siRNA) that trigger the degradation of any other RNA in the cell with a matching sequence. This stops the production of the gene it codes.

Huntington's disease is caused by mutations in one such gene which form clumps of protein that gradually kill off parts of the brain. If a method can be devised to reduce the amount of the toxic protein then this could make significant progress towards a cure for Huntington's disease.[22] A similar approach can be used for other genetic disorders, but care must be exercised as there has been some evidence that introducing siRNA for some disorders might trigger cancer if wrongly switched on.[23]

Gene therapy has great potential use for engineering a virus that seeks out and attacks cancerous cells and tumours anywhere in the body. At present they are not very discriminating and attack any cells unless the dosage is slowly administered at the exact site of the tumour.[24]

An interesting area of research shows that some enzymes scan the DNA chains to locate any damage. It then somehow moves along the chain and repairs these errors.[25] This damage is usually caused by free radicals searching for needed electrons. If the damage is not repaired then the errors are replicated and can be genetically harmful. Working out the mechanism of how the enzymes repair the damage could lead to the scientists themselves being able to make the repairs.

14.7   Some changes of use of existing drugs

The use of existing drugs whether synthetic or natural, in new or novel ways is proving interesting. This includes the drug clioquinol (Figure 14.10), normally used

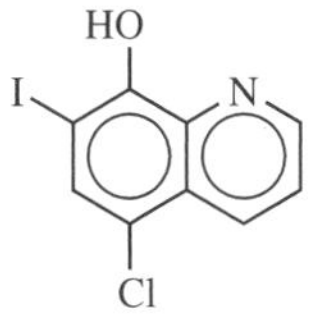

Figure 14.10   Clioquinol

for treating athlete's foot, which has been reported[26] to arrest the symptoms of Alzheimer's disease. It seems to act as an absorber of zinc and copper atoms that accumulate in the brain cells of sufferers of Alzheimer's disease.

Honey has for many years been thought to have healing properties, and a new slant on its use has shown that it can also attack harmful bacteria, including *E. coli*, *Salmonella* and *Helicobacter*, and importantly some anti-bacterial-resistant strains of MRSA.[27] Honey contains hydrogen peroxide, which is harmful to bacteria (hospitals used to use it as a disinfectant at one time). An article in *New Scientist*[28] reported that 'Going to hospital these days has become a bit of a lottery. While your doctors are doing their best to solve the problem that took you there in the first place, nasty hospital bacteria are waiting to ambush you, and perhaps leave you sicker than ever. That happens to 2 million people in the US alone ... Clearly antibiotic resistance is a serious threat'. One of the answers suggested in this article is to produce chemicals that prevent the bacteria sticking to the mucosal cells of our respiratory, gastrointestinal and urinary tracts. The use of anti-adhesive materials could be one answer, some from natural sources others from specially designed molecules. An example is the use of natural cranberry juice, containing natural anti-adhesive chemicals, by women to ward off urinary infections. Chewing gum containing xylitol also reduces throat and ear infections. The polyphenol oxidase, found in potatoes and apples, also prevents the adhesion of many bacteria. So the provreb 'eat an apple a day to keep the doctor away' could have some significance after all and is not just a jingle quoted by grandmothers.

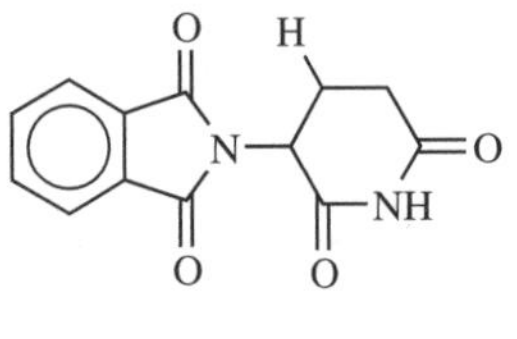

Figure 14.11   Thalidomide

The bad guy of chemical drugs of the 1960s was thalidomide (Figure 14.11) which when taken by women during the first 3 months of pregnancy caused gross limb deformities in their offspring. It was removed from treatment of morning sickness in 1962. It was, however, used in Jerusalem in 1965 as a treatment for gross lesions in cases of leprosy and, amazingly, these lesions rapidly disappeared. Dr Richard Powell, at Nottingham University Hospital, also found that thalidomide was excellent for the treatment of Bechet's disease, which causes painful mouth and congenital ulcers [as reported in a BBC Horizon (MMIV) programme about thalidomide, written by Jill Marshall]. Probably the most amazing use in the

1990s was for treatment for cancers, including blood cancer. The drug attacks the blood supply that feeds the cancers and then kick-starts the immune system.

The problem with thalidomide is that no one knows exactly how and why it works, which makes it difficult to tailor-make a replacement drug. One approach replaced one of the oxygen atoms of the molecule with a nitrogen atom and the new analogue, called Revamid, has been shown in trials to be effective and to have few side effects. This drug is not a cure, but a very effective controller of some cancers. Thalidomide is undergoing a resurrection and many trials and much research are being undertaken.

One of the more recent drugs to become well known is viagra. It was originally developed as a blood-pressure-lowering drug and not to ease erectile dysfunction. Recent uses have been for primary pulmonary hypertension (a progressive narrowing of the arteries in lungs), some cases of heart and lung transplants, babies with high lung pressure and Raynaud's disease or white fingers.

It was widely reported in 2003 that a 'four in one' tablet could help reduce the occurrence of heart failure in middle-aged people by administering it to anyone over 55. It contains aspirin, a statin drug (to help reduce cholesterol), folic acid and a blood pressure controller. However, after taking this 'magic' drug mixture, people would probably feel that they were now safe to smoke and drink more and so negate the effect of the drug. Time will tell.

References

1. G. Cragg and D. Newman. Nature's bounty. *Chemistry in Britain*, January 2001, 22ff.
2. S. Jourdier. A miracle drug. 100 years of Aspirin. *Chemistry in Britain*, February 1999, 33–35.
3. T.M. Brown, A.T. Dronsfield, P.M. Ellis and J.S. Parker. Aspirin – how does it know where to go? *Education in Chemistry*, March 1998, 47–49.
4. *Daily Telegraph*, 20 September 2002, 25.
5. R. Highfield. Search for superbug cure in rock pool slime. *Daily Telegraph*, 27 February 2003, 13.
6. A. Lord. Sweet healing. *New Scientist*, 7 October 2000, 32.
7. R. Creasy. Garlic cure hailed as a breakthrough in killer bug battle. *Sunday Express*, 11 January 2004, 37.
8. P. Jenkins. Taxol branches out. *Chemistry in Britain*, November 1996, 43–46.
9. P.C. McGowan. Cancer chemotherapy gets heavy. *Education in Chemistry*, September 2001, **38**(5), 134–136.
10. D. Derbyshire. *Daily Telegraph*, 1 September 2002, science correspondence notes.
11. Attacking cancer with a light sabre. *Chemistry in Britain*, July 1999, 17.
12. Tamoxifen. Soundbite by Simon Cotton. *Education in Chemistry*, March 2004, **41**(2), 32.
13. G. Hamilton. Hit cancer where it hurts. *New Scientist*, 3 July 2004, 40–43.

14. R. Orwant. Cancer unplugged. *New Scientist*, 14 August 2004, 34–37.

15. B. Austen and M. Manca. Proteins on the brain. *Chemistry in Britain*, January 2000, 28–31.

16. S. Bhattacharya. A balm to soothe troubled minds. *New Scientist*, 28 June 2003, 20.

17. A.T. Dronsfield, M. Hill and J. Pring. Halothane–the first designer anaesthetic. *Education in Chemistry*, September 2002, **39**(5), 131–133.

18. N. Mather. Time to attack. *Education in Chemistry*, March 2000, **62**(InfoChem), 2–3.

19. P. Cohen. Mutator protein helps our bodies fight viruses. *New Scientist*, 28 June 2003, 21.

20. D. MacKenzie. Friend or foe. *New Scientist*, 6 September 2003, 36–37; see also www.newscientist.com/hottopics/sars

21. G. Samuels. Presidential address to the Association for Science Education in 2003, *School Science Review*, June 2003, **84**(309) 31ff.

22. B. Holmes. Switching off Huntington's. *New Scientist*, 15 March 2003, 20.

23. B. Holmes. Cancer risk clouds miracle gene cures. *New Scientist*, 15 March 2003, 6.

24. B. Holmes. Smart virus hunts down tumours wherever they are. *New Scientist*, 15 March 2003, 21.

25. A. Ananthaswamy. Enzymes scan DNA using electric pulse. *New Scientis*, 18 October 2003, 10.

26. M. Day and M. Halle. Drug used to treat athlete's foot slows down Alzheimer's. *Sunday Telegraph*, 11 January 2004, 8.

27. A. Lord. Sweet healing. *New Scientist*, 7 October 2000, 32.

28. A. Ananthaswamy. Taming the beast. *New Scientist* , 29 November 2003, 34–37.

A further more advanced book on these topics is G. Thomas, *Medicinal Chemistry*, 2000, Wiley, Chichester.

15 Numbers and Quantities

Learning objectives

- To give an overview of some relevant units and numerical terms used in chemistry.

Diagnostic test

Try this short test. If you score more than 80 % you can use the chapter as a revision of your knowledge. If you score less than 80 % you probably need to work through the text and test yourself again at the end using the same test. If you still score less than 80 %, come back to the chapter after a few days and read it again.

1. What is a mole? (1)

2. How many grams of a material of molar mass 150 are there in 1 l of 0.001 moles of its solution in water. (2)

3. Which is the more concentrated, 0.02 M solution or 2×10^{-1} M solution. (1)

4. When 100 ml of 0.1 M hydrochloric acid are neutralized by 0.01 M sodium hydroxide to form salt and water, how many millilitres of the alkali are needed? (2)

Chemistry: An Introduction for Medical and Health Sciences, A. Jones
© 2005 John Wiley & Sons, Ltd

5. You have $100 \, cm^3$ of 0.1 M salt solution, NaCl. Express this
   concentration in ppm (parts of salt per million in solution).   (2)

6. What is the percentage of salt (NaCl) in 10 ml of 1 M solution?   (2)

Total 10 (80 % = 8)
Answers at the end of the chapter.
Units, $1 \, cm^3 = 1$ ml; $1 \, dm^3 = 1000$ ml $= 1$ l.

Alice in Wonderland
The Mock Turtle, talking about his education, said:
'I took the regular course.'
'What was that?,' inquired Alice.
'Reeling and Writing and the different branches of Arithmetic – Ambition,
Distraction, Uglification and Derision . . . we also did Mystery, ancient and modern
with Seaography and Drawling . . . And with a classic master we did Laughing and
Grief.'

In this text we will study . . . moles, ancient and modern; Concentrations, weak and
strong; ppm and Dilution, all with a standard notation, with a touch of Mystery and
Uglification, a little Laughing and a lot of Grief!

15.1 Standard notation, powers of 10

The measurement of quantities in science have been standardized, although some older texts and lecturers still use old notations. We have included both in the text where appropriate.

15.1.1 Numbers and multiples of 10 in common use

1000 or 10^3 = kilo, as in kilogram (kg)
$0.001 = 10^{-3}$ = milli, as in milligram (mg) or millimolar (mM)
$0.000001 = 10^{-6}$ = micro, as in microgram (μg)
$0.000000001 = 10^{-9}$ = nano, as in nanogram (ng)
10^{-12} = pico, as in picogram (pg)

For volume the following units are widely used:
$1 \, dm^3 = 1000 \, cm^3$ (sometimes called cc, cubic centimetres) = one litre (1 l)
An older unit sometimes used is litre (l) = 1000 ml
$1 \, cm^3 = 1cc = 1 \, ml = 1 \, mL$. All these mean the same thing but in different nomenclature.
For mass:
1 kilogram (kg) = 1000 g
For length
10^3 m = 1000 m = 1 kilometre (km)
10^{-1} m = 0.1 m = one-tenth of a metre = 1 decimetre (dm)
10^{-2} m = 0.01 m = one-hundredeth of a metre = 1 centimetre (cm)
10^{-3} m = 0.001 m = one-thousandth of a metre = 1 millimetre (mm)
10^{-6} m = 0.000001 = 1 micrometre (μm)
10^{-9} m = 0.000000001 = 1 nanometre (nm)

15.2 Moles

A mole is a made-up unit to describe such a huge number that it is difficult to write in text in ordinary notation with all its zeros. A mole was derived to try to overcome the difficulty of imagining such large numbers. A mole of particles is

600 000 000 000 000 000 000 000

or 6×10^{23} of them. The term is used when talking about the number of particles, atoms or molecules present in a material. You can have a mole of anything as it has no units of its own, so you can have a mole of oranges or of people. A mole has no units because it is a number, just like '12' is a number – you can have 12 oranges or 12 people.

Is there a more simple way of writing it? What about just words? Well, 1000 is a thousand, and a thousand thousand is a million. After that it gets a bit messy, because in the USA a thousand million is a billion, but in the UK (until fairly recently) a million million was a billion. In any case, words take up more space and cause more confusion than numbers do, so it is best to use the more simple notation of powers of 10.

15.3   Powers of numbers and logs

As always, in mathematics and science, we look for the clearest notation – the clearest way of writing things. A square has sides of equal length. Its area is equal to its length × its breadth. Suppose we have a square, each of the sides being '1'; then its area $= 1 \times 1 = 1$. Suppose we have a square, each of the sides being '2'; then its area $= 2 \times 2 = 4$. Similarly, if we have a square, each of the sides being '12', then its area $= 12 \times 12 = 144$. All of this is based on multiplying a number by itself to find the area of a square:

$$3 \times 3 = 9$$
$$5 \times 5 = 25$$
$$7 \times 7 = 49$$

Now, it takes longer to write '7×7' than '7^2'; '7^2' is called 'seven squared', because it relates to the area of a square. Notice the notation: '7^2' means 7 multiplied by another $7 = 7 \times 7 = 49$.

This is a very useful notation. Let us extend it – what is 7^3? The logical extension is that it relates to a cube. A cube has equal sides. Its volume is length × breadth × height, so the volume of the cube $= 7 \times 7 \times 7 = 343$ and 7^3 is called 'seven cubed'.

If $3^3 =$ 'three cubed' $= 3 \times 3 \times 3 = 27$, then 7^4 must mean '$7 \times 7 \times 7 \times 7$' $= 2401$, i.e. four sevens multiplied by each other. What are we going to call this, since '7^4' has no relationship to a real object? We know about squares and cubes – squares have equal sides and are flat – cubes have equal sides and occupy space. There is no name with a real object for '7^4', so it is called 'seven to the power 4' or 'seven to the fourth'.

We (for the most part) have 10 digits and our numbering system is based on 10:

$10^1 = 10$

$10^2 = 10 \times 10 = 100$ ('10' squared = '100')

$10^3 = 10 \times 10 \times 10 = 1000$ ('10' cubed = '1000')

$10^4 = 10 \times 10 \times 10 \times 10 = $ ('10 to the power 4' or '10 to the fourth' = 10 000)

By this time you might have noticed that the power to which 10 is raised on the left-hand side is the same as the number of zeros in the answer.

15.3.1 Logarithms (logs)

The common logs use a number base of 10. The log of any number is the power to which 10 must be raised to equal the number. For example, the common log of 1000 is 3 because 10 must be raised to a power of 3 to make 1000.

$$\log 1000 = \log 10 + \log 10 + \log 10 = 3 \times \log 10 = 3 \times 1 = 3$$

Remember or look up in tables or use a calculator to find that log of $10 = 1$, and the log of $1 = 0$.

The log of 1000 is easy to see but the log of 35 requires the use of a calculator or book of logs.

$$\log 35$$
$$10^x = 35. \text{ What must be the value of } x?$$

Well 10^1 would be 10 and 10^2 would be 100 so the log of 35 must be somewhere between 1 and 2. The tables tell us it is 1.554.

Logs were derived to make the multiplication of large and awkward numbers easier by the process of addition. Taking a simple example

$$1000 \times 250 \times 65$$
$$\log(1000 \times 250 \times 65) = \log 1000 + \log 250 + \log 65$$

Reading from tables or a calculator,

$$3 + 2.3979 + 1.8129 = 7.2108$$

Now look up the tables to take the anti-log of 7.2108:

$$= 16\,250\,000$$

Logs to the base 10 are used to express the concentration of weak acid and alkaline solutions as a more easily understood whole number, pH (see Chapter 9).

$$pH = -\log_{10}[H^+]$$

What is the pH of a solution of a weak acid that has a hydrogen ion concentration of 0.001 M or 1×10^{-3} g/dm^3.

$$
\begin{aligned}
pH &= -\log_{10}[H^+] \\
&= -\log[1 \times 10^{-3}] \\
&= -\{\log 1 + \log 10^{-3}\} \\
&= -\{\log 1 - 3\log 10\} \\
&= -\{0 - 3\} \\
&= 3
\end{aligned}
$$

15.3.2 To go back to the mole

The word 'mole' is a number word – it means

$$600\,000\,000\,000\,000\,000\,000\,000$$

There are 23 zeros in this number. It could be represented as six lots of

$$100\,000\,000\,000\,000\,000\,000\,000$$

or

$$6 \times 100\,000\,000\,000\,000\,000\,000\,000 \text{ or } 600\,000\,000\,000\,000\,000\,000\,000$$

or 6×10^{23} (because there are 23 zeroes after the '6'). This last way is the most abbreviated form and the easiest to write. You will often come across this standard notation of powers of numbers.

$$3 \times 10^2 = \text{is what?}$$
$$3 \times (10 \times 10) = 3 \times (100) = 300$$
$$3.12 \times 10^3 = \text{is what?}$$

$3.12 \times (10 \times 10 \times 10) = 3.12 \times 1000 = 3120$ (three thousand one hundred and twenty)

For the most part, in science it is best to use numbers in standard notation, 3.12×10^3. Usually the number in front of the powers of 10 is kept to only one number before the decimal point.

15.3.3 What about numbers smaller than 1?

Just to be perverse, let us start with 10^3 – a number much bigger than 1 – in fact it means $10 \times 10 \times 10$ or 1000 (or one thousand, if you absolutely have to use language). To get from 10^3 to 10^2, you have to divide by 10. To get from 10^2 to 10^1, you have to divide by 10. So what does 10^1 mean? 10^1 must mean simply 10.

15.3.4 To go a step further

To go from 10^1 to 10^0 you have to divide by 10. However, 10^1 is simply 10, so $10^0 = 10^1$ divided by 10. Therefore 10 divided by $10 = 1$.

$$10^0 = 1$$

Let us divide again by 10. You will (I hope) by now have noticed that each time you divide by 10, the power of 10 goes down by 1:

$$10^1 \text{ divided by } 10 = 10^0 = 1$$

So if we divide 10^0 by 10, the power of 10 must go down by 1 again:

$$10^0 \text{ divided by } 10 = 10^{-1}.$$

What exactly does '10^{-1}' mean? '10^0' is 1, so '10^{-1}' must mean ' 1 divided by 10'. In decimal notation, this is written as '0.1'.

Let us divide by 10 again. We now know that every time we divide by 10, the power of 10 goes down by 1. Thus 10^{-1} divided by $10 = 10^{-2}$; 10^{-2} written in decimal notation as '0.01'.

In the same way, '10^{-3}' must mean '1 divided by 10^3' or '1 divided by 1000', which is written in decimal notation as '0.001'.

It will not have escaped your notice that the number of places after the decimal point to make the number '1' is the same as the negative power of 10.

Thus

$$10^{-7} = 0.000\,000\,1$$

and

$$10^{-12} = 0.000\,000\,000\,001$$

In medicine, small and large numbers are often required, and standard notation is almost always used.

$6 \times 10^8 = 600\,000\,000$ (eight zeros)

$6 \times 10^{-8} = 0.000\,000\,06$ (eight figures after the decimal point to make the number 6)

15.4   Moles in formulae and equations

We know that one mole of any particles, atoms, molecules or oranges would contain 6×10^{23} particles. The mole has been chosen as a unit because, *one mole of atoms has a mass equal to their atomic mass in grams*, e.g. the atomic mass of carbon is 12, so one mole of carbon $= 12$ g.

One mole of molecules will have a mass equivalent to their molecular mass in grams. One mole of sodium chloride, NaCl (Na $= 23$, Cl $= 35.5$) $= 58.5$ g, and one mole of alcohol, C_2H_5OH (C $= 12$, O $= 16$, H $= 1$) $= 46$ g. These 46 g of alcohol will contain the same number of particles as 58.5 g of sodium chloride or 12 g of carbon.

15.4.1   Moles and equations

One mole of any atom or molecule will have a mass of its atomic or molecular mass in grams. This means that 12 g carbon will contain the same number of atoms as there are molecules in 58.5 g sodium chloride.

This is how we know how to balance chemical equations and work out the masses reacting together, because we know that equal molar masses contain the same numbers of particles. In a chemical equation like $C + O_2 \rightarrow CO_2$, we always know that there will be one mole of carbon atoms reacting with one mole of oxygen molecules (O_2) to give one mole of carbon dioxide. Providing the chemical equation balances, any fraction of moles will also be true. One-tenth of a mole of carbon will require one-tenth of a mole of oxygen gas molecules (O_2), to make one-tenth of a mole of carbon dioxide gas (CO_2).

Therefore 10 moles of alcohol have a mass of 460 g and likewise 10^{-2} moles of alcohol have a mass of $46 \times 10^{-2} = 0.46$ g $= 0.46$ g.

15.5 Moles in solution

Molarity is defined as the number of moles of material dissolved in $1\,dm^3$ of solution, i.e. mol/dm^3 or in older notation mol/l. The molarity of a solution = number of grams of solute divided by its molecular mass (i.e. the number of moles) in $1\,dm^3$, or 1 l, of solution.

When a substance is soluble in water (or any other solvent also applies), we have a standard definition that says that 'One litre (dm^3) of solution that contains one mole of a solute is called 1 M.'

$1\,dm^3$ (litre) of 1 M salt solution contains 58.5 g

or $1000\,cm^3$ of 1 M salt solution contains 58.5 g

1 l or $1000\,cm^3$ of 0.1 M (or 10^{-1} M) salt solution contains 5.85 g

$100\,cm^3$ of 0.1 M salt solution contains 0.585 g

$1\,cm^3$ of 0.1 M salt solution contains 0.00585 g or 5.58×10^{-3} g

Try the following using the same logic as above: how many grams of glucose ($C_6H_{12}O_6$), C=12, H=1, O=16, molar mass=180) will it take to make the following solutions:

1. 1 l of 1 M solution;

2. $1000\,cm^3$ of 0.1 M solution;

3. $100\,cm^3$ of 0.1 M solution;

4. $1\,cm^3$ of 0.1 M solution;

5. $100\,cm^3$ of a 0.2 M solution?

Expressing the concentration of solutions in terms of moles is the most common method:

1 M = 1 molar

0.1 M = decimolar

0.01 M = centimolar

0.001 M = millimolar

Again, we know that equi-molar solutions will contain the same number of moles. That is how we know we can balance equations and trust our calculations of concentrations of solutions.

Quantities of solids are usually expressed in the number of grams of the material rather than the number of moles. The abbreviation for a gram is g:

0.1 g = 1 decigram or 10^{-1} g

0.01 g = 1 centigram or 10^{-2} g

$$0.001 \text{ g} = 1 \text{ milligram (mg) or } 10^{-3} \text{ g}$$
$$0.000001 \text{ g} = 1 \text{ microgram (}\mu\text{g) or } 10^{-6} \text{ g}$$
$$10^{-9} \text{ g} = 1 \text{ nanogram (ng)}$$
$$1000 \text{ g} = 10^{3} \text{ g} = 1 \text{ kilogram (kg)}$$

15.6   Concentration in ppm, parts per million

This notation is based on the masses (in grams) of materials usually in very very dilute aqueous solutions where the density of the solution is virtually that of the water solvent, i.e. 1 g/cm^3.

$$1 \text{ cm}^3 \text{ of solution} = 1 \text{ g}$$

It means the number of parts (mass) in one million parts (mass) of solution.

Therefore a solution containing 0.001 g of solute in 1000 cm^3 is 1 ppm and could be worked out as follows. Remembering that 1 cm^3 of weak solution will have a mass of 1 g, 1000 cm^3 is 10^3 cm^3 and contains 0.001 g or 1×10^{-3} g of solute, or expressing it another way, 1×10^{-3} g of solute in 1000 cm^3 of solution, so multiplying by 1000 would mean 1 g of solid would be contained in 1000×1000 cm^3 of solution, which is 1 g in 1 000 000 cm^3 (or g), which is 1 g in 1 000 000 g solution or 1 g per million = 1 ppm. Therefore, one milligram (0.001 g, 10^{-3} g) of a material in 1 dm^3 (1000 cm^3) is equivalent to 1 ppm.

In some American texts the concentrations are expressed as parts per billion. The word billion is the American definition of billion as 1000 million not the more frequently used British billion of a million million. Great care must be taken to know which units are being used because in the UK many people also use the USA definition of billion. Here 1 μg (10^{-6} g) of material in 1 dm^3 is equivalent to 1 ppb.

15.7   Dilutions

A bottle of concentrated orange squash must be diluted to make it palatable and the quantities are usually estimated. In science the dilution must be accurate as lives might depend upon it. If 1 cm^3 of 1 M salt is diluted with water to make it up to 1000 cm^3, the volume goes up from 1 to 1000 and the concentration goes down by the same proportion, i.e. by 1000. Therefore the diluted solution is now 0.001 M or 1×10^{-3} M or mM.

15.8 Percentage by mass

In a dry powder the percentage of each constituent in a mixture is calculated in grams as expressed as a proportion of the total mass of 100 g of the sample; 100 g of a mixture containing 30 g of substance A and 70 g of B would have a percentage composition of 30 % A and 70 % B.

If a solution is very dilute, the mass of the active materials in water solution can also be expressed as a percentage because the mass of the materials present is negligible compared with the mass of the water.

0.01 g (1×10^{-2}) of salt in 100 cm^3 of solution is therefore 0.01 % salt
20 cm^3 of this 0.01 % salt solution would contain $0.01/5 = 0.002$ g.
so 10 % salt solution is 10 g of salt in 100 cm^3 (100 g) of solution.

We appreciate that the figures in this chapter are mind-bending, but it is essential that correct values are understood when prescribing quantities for patients. It might be useful to make a single sheet of the most used units and keep them handy for reference.

Answers to the diagnostic test

1. The number 6×10^{23} (1)

2. 0.15 g   (2)

3. 2×10^{-1} M   (1)

4. 1000 ml   (2)

5. 5850 g or 58.5×10^2 g per million g of solution

6. 5.85 %   (2)

Answers to questions in the text

How many grams of glucose ($C_6H_{12}O_6$), $C = 12$, $H = 1$, $O = 16$, Molar mass $= 180$) will it take to make the following solutions?

1. $1\,l$ of $1\,M$ solution, $180\,g$;

2. $1000\,cm^3$ of $0.1\,M$ solution, $18\,g$;

3. $100\,cm^3$ of $0.1\,M$ solution, $1.8\,g$;

4. $1\,cm^3$ of $0.1\,M$ solution, $0.18\,g$;

5. $100\,cm^3$ of $0.2\,M$ solution, $3.6\,g$.

Further questions

1. What is the molarity of the following:

    i. $18\,g$ of glucose dissolved in $100\,cm^3$ of solution;
    ii. $10\,\%$ solution of salt in water;
    iii. a solution of 1 part per million of salt solution ($1\,g$ in $10^6\,g$ solution)?

    Atomic masses, $H = 1$, $C = 12$, $O = 16$, $Na = 23$, $Cl = 35.5$.

2. How many cm^3 of $2\,M$ hydrochloric acid solution are needed to be diluted to make up a $1\,l$ ($1\,dm^3$) solution of $0.1\,M$?

3. How many grams of material will be needed to make up the following solutions:

    i. $500\,cm^3$ of $10\,\%$ solution of substance X;
    ii. $200\,cm^3$ of $2\,M$ salt solution;
    iii. $300\,dm^3$ of 6 ppm of salt?
    iv. What is the percentage by mass of hydrogen in water, H_2O.

    Atomic masses of $H = 1$, $O = 16$, $Na = 23$, $Cl = 35.5$.

4. What is the pH of the following solutions:

    i. $0.01\,M$ hydrochloric acid, HCl;
    ii. $0.01\,M$ sodium hydroxide solution, NaOH (remember to use $K_w = 10^{-14}$, $OH^- \times H^+ = 10^{-14}$ or $H^+ = 10^{-14}/OH^-$)?

Answers to further questions

1. i.   1 M;
   ii.   1.7 M;
   iii. 1.7×10^{-5} M.

2. 50 cm^3.

3. i.   50 g;
   ii.   23.4 g;
   iii. 18×10^{-4} g;
   iv.   11.1 %.

4. i.   pH $= 2$;
   ii.   pH $= 12$.

Appendix 1: Alphabetical List of the Common Elements

Name	Symbol	Atomic no.	Atomic mass	Name	Symbol	Atomic no.	Atomic mass
Actinium	Ac	89	227	Fermium	Fm	100	257
Aluminium	Al	13	27.0	Fluorine	F	9	19.0
Americium	Am	95	243	Francium	Fr	87	223
Antimony	Sb	51	121	Gadolinium	Gd	64	157
Argon	Ar	18	40.0	Gallium	Ga	31	69.2
Arsenic	As	33	74.9	Germanium	Ge	32	72.6
Astatine	At	85	210	Gold	Au	79	197
Barium	Ba	56	137	Hafnium	Hf	72	178.5
Berkelium	Bk	97	247	Hassium	Hs	108	265
Beryllium	Be	4	9.0	Helium	He	2	4.0
Bismuth	Bi	83	209	Holmium	Ho	67	165
Bohrium	Bh	107	264	Hydrogen	H	1	1.0
Boron	B	5	10.9	Indium	In	49	115
Bromine	Br	35	79.9	Iodine	I	53	127
Cadmium	Cd	48	112	Iridium	Ir	77	192
Caesium	Cs	55	133	Iron	Fe	26	55.8
Calcium	Ca	20	40.1	Krypton	Kr	36	83.8
Californium	Cf	98	251	Lanthanum	La	57	139
Carbon	C	6	12.0	Lawrencium	Lw	103	260
Cerium	Ce	58	140	Lead	Pb	82	207
Chlorine	Cl	17	35.5	Lithium	Li	3	6.9
Chromium	Cr	24	52.0	Lutetium	Lu	71	175
Cobalt	Co	27	58.9	Magnesium	Mg	12	24.3
Copper	Cu	29	63.5	Manganese	Mn	25	54.9
Curium	Cu	96	247	Meitnerium	Mt	109	266
Dubnium	Db	105	262	Mendelevium	Md	101	258
Dysprosium	Dy	66	162.5	Mercury	Hg	80	200
Einsteinium	Es	99	254	Molybdenum	Mo	42	95.5
Erbium	Er	68	167	Neodymium	Nd	60	144
Europium	Eu	63	152	Neon	Ne	10	20.2

Chemistry: An Introduction for Medical and Health Sciences, A. Jones
© 2005 John Wiley & Sons, Ltd

Name	Symbol	Atomic no.	Atomic mass	Name	Symbol	Atomic no.	Atomic mass
Neptunium	Np	93	237	Seaborgium	Sg	106	263
Nickel	Ni	28	58.7	Selenium	Se	34	79.0
Niobium	Nb	41	92.9	Silicon	Si	14	28.1
Nitrogen	N	7	14.0	Silver	Ag	47	108
Nobelium	No	102	259	Sodium	Na	11	23.0
Osmium	Os	76	190	Strontium	Sr	38	87.6
Oxygen	O	8	16.0	Sulfur (Sulphur)	S	16	32.1
Palladium	Pd	46	106	Tantalum	Ta	73	181
Phosphorus	P	15	31	Technetium	Tc	43	99.0
Platinium	Pt	78	195	Tellurium	Te	52	128
Plutonium	Pu	94	244	Terbium	Tb	65	159
Polonium	Po	84	209	Thallium	Tl	81	204
Potassium	K	19	39.1	Thorium	Th	90	232
Praseodymium	Pr	59	141	Thulium	Tm	69	169
Promethium	Pm	61	145	Tin	Sn	50	119
Protactinium	Pa	91	231	Titanium	Ti	22	47.9
Radium	Ra	88	226	Tungsten	W	74	184
Radon	Rn	86	222	Uranium	U	92	238
Rhenium	Re	75	186	Vanadium	V	23	50.9
Rhodium	Rh	45	103	Xenon	Xe	54	131
Rubidium	Rb	37	85.5	Ytterbium	Yb	70	173
Ruthenium	Ru	44	101	Yttrium	Y	39	88.9
Rutherfordium	Rf	104	261	Zinc	Zn	30	65.4
Samarium	Sm	62	150	Zirconium	Zr	40	91.2
Scandium	Sc	21	45.0				

Elements above 103 have been made artificially up to element 116, but they exist for only a matter of microseconds or less. Some have not been officially named as yet.

Appendix 2: Periodic Classification of the Common Elements

According to their atomic numbers (top line)
and their approximate atomic masses (bottom line)

1 H 1																	2 He 4
3 Li 7	4 Be 9											5 B 11	6 C 12	7 N 14	8 O 16	9 F 19	10 Ne 20
11 Na 23	12 Mg 24											13 Al 27	14 Si 28	15 P 31	16 S 32	17 Cl 35	18 Ar 40
19 K 39	20 Ca 40	21 Sc 45	22 Ti 48	23 V 51	24 Cr 52	25 Mn 55	26 Fe 56	27 Co 59	28 Ni 59	29 Cu 64	30 Zn 65	31 Ga 70	32 Ge 72	33 As 75	34 Se 79	35 Br 80	36 Kr 84
37 Rb 85	38 Sr 88	39 Y 89	40 Zr 91	41 Nb 93	42 Mo 96	43 Tc 99	44 Ru 101	45 Rh 103	46 Pd 106	47 Ag 108	48 Cd 112	49 In 115	50 Sn 119	51 Sb 122	52 Te 128	53 I 127	54 Xe 131
55 Cs 133	56 Ba 137	57 La 139	72 Hf 179	73 Ta 181	74 W 184	75 Re 186	76 Os 190	77 Ir 192	78 Pt 195	79 Au 197	80 Hg 201	81 Tl 204	82 Pb 207	83 Bi 209	84 Po 209	85 At 210	86 Rn 222
87 Fr 223	88 Ra 226	89 Ac 227	104 Rf 261	105 Db 262	106 Sg 263	107 Bh 262	108 Hs 265	109 Mt 266									

Lanthanum Series

58 Ce 140	59 Pr 141	60 Nd 144	61 Pm 145	62 Sm 150	63 Eu 152	64 Gd 157	65 Tb 159	66 Dy 163	67 Ho 165	68 Er 167	69 Tm 169	70 Yb 173	71 Lu 175

Actinium Series

90 Th 232	91 Pa 231	92 U 238	93 Np 237	94 Pu 244	95 Am 243	96 Cm 247	97 Bk 247	98 Cf 251	99 Es 254	100 Fm 257	101 Md 257	102 No 259	103 Lr 260

Elements above 103 have been made artificially up to element 116, but they exist for only a matter of microseconds or less. Some have not been officially named as yet and not clearly assigned to positions in the table.

Glossary

Acetylcholine
A neurotransmitter that is stored in vesicles at the nerve endings and secreted when a message needs to be passed from one nerve to another alongside it.

Acid
A substance that produces hydrogen ions, H^+, when dissolved in water.

Adenosine diphosphate, ADP
A compound found in the cells and involved in the storage of energy. It combines with a further phosphate group to form ATP or adenosine triphosphate, which is a high-energy compound that is essential for cell energetics.

Aerobic
Processes that require the presence of air or oxygen.

AIDS
A disease that develops when HIV overcomes the immune system of the body.

Alcohol
An organic compound that contains at least one OH group.

Aldehyde
An organic compound that contains a HC=O group.

Alkali
A substance that produces hydroxyl ions, OH^-, when dissolved in water. An alkali is a soluble base. A base is usually an oxide or hydroxide of a metal.

Alkaloids
A large group of nitrogen bases found in seeds, berries and roots of plants. Many are poisonous but others have found a use in medicine, including morphine. Caffeine is also an alkaloid but is not poisonous in small amounts.

Alkane
An organic compound that contains only carbon and hydrogen atoms. There are many thousands of these compounds arranged in a homologous series having the general formula C_nH_{2n+2}. When $n = 1$ the alkane is CH_4, methane.

Alkyl group
The part of an alkane with one hydrogen removed; a methyl group is CH_3 from CH_4.

Alpha particle
A positively charged particle emitted from the nucleus of some radioactive atoms. It is made up of two neutrons and two protons and no electrons. It is shown as the nucleus of a helium atom, $^4_2He^{2+}$.

Amine
An organic compound with similar alkaline properties to ammonia. General formula $C_nH_{2n+1}NH_2$, e.g. ethylamine, when $n = 2$, $C_2H_5NH_2$.

Chemistry: An Introduction for Medical and Health Sciences, A. Jones
© 2005 John Wiley & Sons, Ltd

Amino acid   An organic compound with both amine and carboxylic acid groups in it. It has both alkaline and acid properties and is capable of forming a zwitter ion. This has both a positive and negative charge at either end of the molecule. They are the building blocks of proteins, e.g. $^+H_3N \cdot CH_2 \cdot COO^-$ a zwitter ion of glycine; $H_2N \cdot CH_2 \cdot COOH$, the simplest of the amino acid series.

Anabolic   A metabolic process that builds up small units into larger ones, e.g. amino acids into peptides and proteins inside the cells.

Anaerobic   Processes that occur without the presence of air or oxygen

Analysis, analyse   A method used to determine the properties or composition of a compound.

Anode and anions   The positive electrode of an electrical cell. They attract anions (negative ions), e.g. Cl^- ions.

Antibiotic   Chemicals made by bacteria or fungi. Penicillin is such a compound.

Antiseptic   Chemicals which kill microorganisms.

Aqueous solution   A solution of a material in water.

Aromatic   An organic compound containing a ring structure with alternate single and double bonds, e.g. benzene, C_6H_6.

Atom   The smallest particle of an element that shows all the properties of that element.

Atomic mass and relative atomic mass   The number of times the mass of an atom of an element is heavier than one atom of hydrogen or one-twelfth the mass of the 12 isotope of carbon.

Atomic number   The number of positively charged particles (protons) in the nucleus of an atom.

Atomic structure   The structure of an atom showing all the particles on the nucleus and also the electrons orbiting around it.

Balancing chemical equations   A chemical equation is simply a way of representing what happens when substances react together. It can be in the form of a word equation or it can be represented by chemical symbols, this being the 'short-hand' way of saying what is happening. There must always be equal amounts of atoms or ions on each side of the equation. Take for example the chemical reaction between hydrogen gas, H_2, and oxygen gas, O_2. The word equation would be hydrogen + oxygen → water. Note that this is a qualitative description and does not tell you how much is reacting. However, the symbol equation is more quantitative as it indicates the numbers of moles reacting, $2H_2 + O_2 \rightarrow 2 H_2O$. Note that there are the same numbers of atoms on each side of the equation. The formulae of the individual molecules must not change; to balance an equation only the numbers in front of the molecules can be changed. The equation is really saying that two moles of hydrogen gas are reacting with one mole of oxygen gas to give two moles of water. Notice that the equations are only part of the story and do not tell you anything about conditions, temperatures etc., but the conditions can sometimes be included in an equation if so required. It would then be $2H_2(g) + O_2(g) \rightarrow 2 H_2O(l)$. This shows gases (g) and liquids (l); a solid is shown by the inclusion of (s) after its symbol. Balanced equations are useful in calculating quantities reacting together, because the equation $2H_2 + O_2 \rightarrow 2\,H_2O$ also says 2 moles of hydrogen gas each weighing 2 g, i.e. $2 \times (1)_2 = 4$ g react with one mole of $(16)_2 = 32$ g of oxygen, to give two moles of water each weighing $(1)_2 + 16 = 18$, i.e. a total of 36 g.

Base
(a) A compound that neutralizes an acid. It is usually a metal oxide. Those bases that are soluble in water are called alkalis and produce OH^- ions when dissolved in water.

(b) In organic chemistry, a compound that contains nitrogen atoms, usually as an amine group. The term 'base' is also used for a group of compounds that are involved with the structure of DNA and RNA. These include thymine, cytosine, guanine and adenine. In these compounds they are usually attached to a sugar.

Becquerel
A unit of measurement of radioactivity; 1 Bq represents one disintegration per second. A larger unit is the Curie (Ci), which is 3.7×10^{10} disintegrations per second. This is the number of disintegrations emitted from 1 g radium.

Beta blocker
A chemical that acts on the β-andrenergic receptors in the myocardium and peripheral vessels.

Beta particle
A fast-moving electron emitted from the nucleus of a radioactive element. It is a small particle with a range of about 20 cm.

Binary compound
A compound formed of two different atoms, e.g. carbon and oxygen in carbon dioxide, CO_2.

Biochemistry
The study of compounds found in living systems.

Boiling point
The temperature at which a liquid boils.

Bond or chemical bond
The electron arrangement holding two atoms together.

Buffer solution
A solution that maintains a constant pH even when impurities are added.

Burning
A chemical reaction with air or oxygen that produces heat and light.

Calorie
The heat required to raise 1 g water by 1 °C. A kilocalorie (kcal) is the quantity of heat needed to raise 1 kg water by 1 °C. It is used as an older unit of energy but has been replaced by the joule; 1 calorie $= 4.184$ J.

Carbohydrate
A class of compounds containing carbon, hydrogen and oxygen in which the ratio of hydrogen to oxygen is 2:1, as in water (H_2O). Glucose is a carbohydrate, $C_6H_{12}O_6$.

Carbonyl group
A compound containing the C=O group. Aldehydes, ketone, sugars and proteins all contain the C=O bond.

Carboxylic acid
An organic molecule containing the —COOH group, e.g. CH_3COOH, ethanoic acid, commonly known as acetic acid.

Catabolism and catabolic
A metabolic process that breaks molecules apart, e.g. breaking down proteins into amino acids.

Catalyst
A substance that alters the rate of a chemical reaction without being used up. Enzymes are biological catalysts.

Cathode
The negative electrode of an electrical cell

Cation   A positive ions attracted to a cathode, e.g. Na^+ ions.

Centrifuge   A machine that spins tubes of liquids at high speed, forcing any heavier materials suspended in the liquid to the bottom of the tube. The process is used for separation of plasma from the heavier blood particles.

Chemical bond   See 'bond'. There are two main types see *covalent bonding* or *ionic bonding*.

Chemical equation   See *balanced chemical equation*.

Chemical formulae   Symbols for elements, see symbols. A method for writing simple formulae for common materials. Each element has its own value of its reacting power and depends upon its position in the periodic table. When elements react together they form chemical bonds (look up *covalent bonding* and *ionic bonding*).

Chemical properties   The properties of a chemical (atoms or molecules) that involve a permanent change to a new substance. For example carbon burns in air to form carbon dioxide.

Chemical reactions   Why do they occur?
                            Chemicals react together in order to reach a lower and more stable energy state. If atoms react together they do so to achieve a stable outer electron shell of all the atoms in the new molecules (see *ionic bonding* and *covalent bonding*).

Chiral   A molecule that contains an optically active group which twists the plane of polarized light. See also *optical activity*.

Cholesterol   A naturally occurring compound in the body but, if it builds up in arteries, particularly those of the heart, it can cause the restriction of the blood flow and bring about heart attacks. It belongs to the hormone family of compounds.

Chromosome   A thread-like strand of DNA in the nucleus of a cell. It contains all the characteristics of the cell in its coded sequence of groups.

Coenzyme   A small molecule that helps an enzyme to work efficiently. These are often metal ions.

Colloid or colloidal   This is a homogeneous mixture of particles of one material uniformly suspended in another. After a long time the heavier particles begin to settle out, e.g. the particle of fat held in a homogenous dispersion in milk or mayonnaise.

Compound   A chemically bound material containing at least two different atoms. The atoms are held by either covalent or ionic bonding within the compound.

Condensation   This has two meanings: (a) it can mean the changing of a vapour to a liquid on cooling, as in the case of steam going to water; (b) it is a term used in organic chemistry to mean two molecules joining together to form a new compound with the elimination of a small molecule (usually a water molecule), e.g. two amino acids condensing to form a dipeptide.

Covalent bonding   This is the method of holding two or more atoms together where each atom shares its electrons with the other so that each atom achieves a stable 'octet' arrangement of its outer electrons.

Curie	A unit of radioactivity which represents 3.7×10^{10} disintegrations per second. It represents the number of disintegrations emitted from 1 g radium.
Decay series	When the nucleus of a radioactive atom disintegrates, it emits various particles and so changes its own composition. When an alpha particle is lost then a new element is formed, which is two places to the left in the periodic table. When a beta particle is lost then a new element is formed which is one place to the right in the periodic table. Therefore, by a series of losses of alpha and beta particles, the element progressively changes. This is called 'decay', and the pattern it follows until a stable nuclear arrangement is reached (usually when the element lead is formed) is called the decay series (see Chapter 12).
Dialysis	This is the purification of a liquid, often a body fluid, using a permeable membrane and applied pressure.
Diffusion	This is another name for mixing. It is usually used when gases mix; they diffuse together. In liquids we usually say they mix or dissolve rather than diffusing. The diffusion is caused by the rapid movement of particles.
Dissociate	When a molecule splits apart, such as when salt, sodium chloride, dissolves in water to form the separate ions of Na^+ and Cl^-. This is also called ionization.
Distillation	This if the continuous process involving evaporation of a liquid into its vapour by heating the liquid. This is followed by cooling and so condensing the vapour back to the liquid. It is used to extract a pure sample of a liquid (or solvent) from an impure mixture, e.g pure water from dirty water or making pure alcohol from a fermented mixture.
DMARD	Disease-modifying anti-rheumatic drugs.
DNA, deoxyribonucleic acid	A massive polymer molecule that contains all the genetic information of the cell. It has a complicated structure containing many thousands of smaller units of amino acids, proteins, bases and sugars. It forms a double-twisted helix string-like molecule. The strands are held in position by cross links of hydrogen bonds from matching pairs of base materials on opposite chains. It is found in the nucleus of cells and it stores genetic information.
Double bond	This is a covalent bond comprising two electrons from each of the sharing atoms. It is most common in organic molecules, such as ethene, $H_2C{=}CH_2$, or in carbon dioxide, $O{=}C{=}O$.
Electrode	The un-reactive material that takes the DC current into a solution of ions. The negative electrode is called the cathode and the positive the anode.
Electrolysis	This is the process of applying a DC current through electrodes to a solution that contains ions. The ions then begin to migrate to the oppositely charged electrodes and are discharged.

Electrolyte	This is a solution containing ions that conduct electricity when applied. It contains both positively and negatively charged particles called ions.
Electromagnetic spectrum	The wide spread of radiation including visible colours (and unseen radiation, like ultraviolet and infrared) that composes the full energy range of radiation.
Electron	A very small particle of negative electricity.
Electronic structure of atoms	The pattern of negative electrons of an atom that orbit around the positively charged nucleus.
Electrophoresis	The movement of ions, usually from biological samples, across wet paper. The ions are attracted to the oppositely charged electrodes. The process is used in analytical laboratories in hospitals to detect what ions are present in a sample.
Electrostatic force	The formation of the forces of attraction between two oppositely charged species, or repulsion between charges of the same type. These charges can gather on the surface of materials like clothing, hair or metal. Lightning is caused by the discharge of these oppositely charged particles between a cloud and the ground.
Electrovalent bonding	See *ionic bonding*.
Elements	The most simple material comprising atoms that still show all the properties of that element. The elements can be put into an ordered pattern called the periodic table. There are some 100 naturally occurring elements and a further 10 that can only be made artificially. See a copy of the Periodic Table.
Enthalpy	A measure of the heat energy needed or produced by a chemical reaction.
Entropy	This measures the molecular disorder of any system. The second law of thermodynamics says that everything is moving in the direction of increasing disorder or increasing entropy.
Enzyme	These are biological molecules that change the rate of chemical reactions in the body. They are biological catalysts, e.g. catalase in the blood decomposes harmful hydrogen peroxide.
Equilibrium	A state whereby the forward process is counterbalanced by a backward reaction.
Esters	A series of compounds made by the interaction of an organic acid and an alcohol.
Ethers	A series of compounds, but usually the word 'ether' refers to only one member of the homologous series, diethyl ether or ethoxy ethane $C_2H_5OC_2H_5$. It is a heavy vapour with anaesthetic properties and was one of the earliest materials used to induce unconsciousness in patients. Better anaesthetics have been developed that are less harmful and easier to use.
Evaporation	When a liquid changes to its vapour it is said to evaporate. The higher the temperature the more a material evaporates.
Exothermic	A chemical reaction that gives out heat, e.g. when a fuel burns in air. The opposite of this process is when a chemical reaction needs heat to react at all, this is called an 'endothermic' chemical reaction.

Fatty acid	A long-chain carboxylic acid that form part of the fats and lipids.
Formula	A symbolic way of showing the constituents of a molecule, e.g. $C_6H_{12}O_6$ as the formula for glucose.
Formula mass	The sum of the masses of all the atoms making up a compound, e.g. $C_6H_{12}O_6 = 6(12) + 12(1) + 6(16) = 180$.
Fractional distillation	The process of boiling off, at its own characteristic boiling point, the different parts or fractions in a mixture of liquids. Each 'fraction' can then be condensed back to a pure separate part of the liquid. It is a process used in separating the different constituents from crude oil to make fuels and separate the petrol from diesel oils and heavy lubrication oils.
Free radical	An atom or group of atoms that has overall spare un-bonded electrons. They are usually short lived and very active species.
Functional group	The part of an organic molecule that determines its characteristic properties, e.g. in alcohols like ethanol, C_2H_5OH, it is the OH that determines its properties.
Gamma radiation	Extremely dangerous penetrating radiation emitted when a radioactive nucleus of an atom decomposes. It penetrates paper, metals and bricks and can only be stopped by a block of lead shielding. It is similar to X rays.
Glucose $(C_6H_{12}O_6)$	A monosaccharide providing the cells with an energy source.
Glycogen	A polysaccharide made up of chains of glucose molecules. It is stored mainly in the liver and muscles. It is an energy store for cells.
Haemoglobin	A large protein with four binding sites for oxygen. It carries oxygen around the body for use by cells. It contains iron in its structure (haem). It also acts as a buffer to maintain a constant working pH of the blood.
Half-life	An expression of the activity of a radioactive isotope. It is the time it takes for radioactivity from an element to fall to half its value. The shorter its half-life the more dangerous a substance is.
Halogen	The elements of group 7 of the periodic table, e.g. chlorine, bromine, iodine.
Heterogeneous	Two materials not in the same physical state.
HIV	Human Immuno-deficiency Virus.
Homogeneous	Two or more substances in the same state, e.g. gin and tonic or whiskey and soda!
Hormone	Chemicals that are present in small quantities in the blood all the time but extra quantities are excreted by glands in response to certain stimuli, such as fear or sexual stimulation, e.g. testosterone and progesterone.
Hydrate and dehydrate	Hydration is the addition of water in a chemical reaction and dehydration is the taking away of water from a substance.
Hydrogenation	The addition of hydrogen gas to a double-bonded 'unsaturated' carbon = carbon bond, making the compound 'saturated' and containing only single C—C bonds.

Hydrogen bond	This is a special type of bond that only occurs between hydrogen atoms and certain electronegative atoms on the right-hand side of the periodic table, namely fluorine, chlorine, oxygen and nitrogen atoms. It is due to a slight difference in charge distributed within the molecule and between hydrogen and these atoms. The hydrogen becomes slightly positively charged as compared with the other atom (say oxygen). This charge then attracts the oppositely charged atoms of another molecule close by, e.g. in water hydrogen bonds keep three H_2O molecules joined together.
Hydrolyse or hydrolysis	This is the chemical reaction of water with another compound, e.g. a protein can be broken down by hydrolysis (H_2O) into the separate amino acids. Here the water attacks the peptide bonds, $-NH \cdot CO \cdot -$ groups.
Hydroxonium ion, H_3O^+	A hydrated proton or a more complete way of showing what a hydrogen ion really looks like: $H^+ + H_2O \rightarrow H_3O^+$.
Inorganic	This term means anything that in not 'organic' (containing carbon atoms). It is often used to describe compounds of metals. They are usually ionic in character.
Insoluble	A substance that does not dissolve in a solvent.
Ionic bonding	The type of bonding between oppositely charged particles held together by the strong attractive forces.
Ionization	A process whereby an ionic molecule, like sodium chloride, NaCl, separates to form the individual ions: $NaCl \rightarrow Na^+ + Cl^-$. This can occur when it dissolves in water or when it melts.
Ions	These are charged particles. Some have positive charges, e.g. Na^+, and are generally called cations. The negative ions, e.g. OH^-, are called anions.
Isomers	A compound that has two or more alternate structures but the same molecular formula. Isomers differ in physical properties from each other. There are many different types of isomerism, e.g. structural, optical, functional group and geometrical isomerism. The term is most widely used in organic chemistry.
Isotonic	These are solutions that exert the same osmotic pressure. It is important to the cells of our body that solutions and fluids are isotonic in order to keep our liquid system in balanced equilibrium.
Isotope	Isotopes have the same atomic number (same number of protons) but differ in the number of neutrons on the nucleus.
Kinetic energy	The energy due the movement of particles. Increasing the temperature causes the particles of the material to move more or vibrate, increasing their kinetic energy.
Kinetics	This is the study of the speed and mechanism of a chemical reaction.
Krebs cycle	A cyclic system and set of chemical pathways within cells to make and use energy. Also known as the citric acid cycle.
Lipid	A 'fatty' organic material found in plants and animal cells. There is a wide range of different types of lipids. They are usually esters of fatty acids and glycerol.

Liquefaction   The changing of a gas to a liquid.

Mass number   The total number of protons and neutrons in the nucleus in an atom.

Melting point   The exact temperature at which a solid changes into its liquid on heating. At the exact melting point both the solid substance and its liquid are in equilibrium. Pure substances have characteristic melting points but impure mixtures melt over a wide range.

Metabolism   The general term used to describe the sum of all the many reactions going on inside a cell.

Metal   They are the most abundant collection of atoms. The metals are usually found on the left-hand side of the periodic table and comprise approximately 80 % of all elements.

Micelle   A large groups of materials formed when a soap or detergent removes grease, dirt or fat from a material and disperses in water. The carbon chain of the detergent molecule dissolves in the grease and its water-soluble head (usually $COONa$ or SO_3Na) helps to disperse it in water. The large detergent/grease groups are called micelles.

Mineral   A naturally occurring compound, usually of a metal. The rocks of the Earth contains these minerals.

Mixture   You can obtain a mixture of any materials but the characteristics of it will be that the separate components can by some means be separated out from one another.

Molarity (M)   This is a way of expressing concentration. One molar (1 M) concentration contains the mass one mole in grams dissolved in $1\,l$ ($1\,dm^3$) of solution.

Molar mass   The mass in grams of a one mole of a substance.

Mole   A general quantity for any substance. A mole of materials, atoms, molecules or particles, is a quantity that contains 6×10^{23} particles. One mole of hydrogen atoms weighs 1 g, one mole of carbon atoms weighs 12 g. It is possible to call 6×10^{23} oranges, one mole of oranges, but I do not think the idea would catch on, not even in Europe!

Molecule   This is the smallest particle of a compound that can exist free in nature and contains at least two atoms joined together by chemical bonds.

Monosaccharide   The class of sugars or carbohydrates that contains one unit of the sugar. It is the simplest class of sugars, e.g. $C_6H_{12}O_6$, glucose.

MRSA   Methicillin-resistant *Staphylococcus aureus* (MRSA). Often called the hospital superbug because it is resistant to most of penicillin treatments.

Neutralization   When an acid is just counteracted by an alkali we say it has been neutralized.

Neutron   A neutral particle that makes up the nucleus of atoms.

Nitric oxide   Contains one atom of nitrogen and one atom of oxygen, NO. It is made and released in small quantities in many body functions. It plays an important role in controlling many systems in our body, and also has a role in controlling the blood flow. In addition, it is a cell signalling molecule to parts of the brain. The complete mechanism of how it works is still being researched.

Noble gas or inert gas
Gases form the group 8 of the periodic table. All contain a complete outer shell of electrons. They are helium, neon, argon and krypton.

Nonmetals
Elements usually at the centre or on the right-hand side of the periodic table, e.g. carbon, nitrogen, oxygen, chlorine. Many of them are gases.

NSAID
Non-steroidal anti-inflammatory drugs, e.g. aspirin.

Nucleon
The term for any particle on the nucleus of an atom.

Nucleus of an atom
Mainly composed of positively charged protons and also neutral neutrons of approximately the same mass. The mass of an atom is almost wholly due to the masses of these particles.

Octet rule
A system of chemical bonding whereby atoms try to obtain a stable outer electron arrangement of a noble gas by achieving a complete octet of electrons in their outer shell. See also *covalent bonding* and *ionic bonding*.

Optical activity
When a molecule forms an asymmetric arrangement such that two isomers can be formed, one being the mirror image of the other. Carbon molecules that contain four different groups attached to a central carbon atom are such compounds. See also *isomers* and *chiral*. The *R/S* system of naming isomers depends on using an order of priority of groups based on their atomic mass which are attached to an asymmetric carbon atom in a molecule. A model of the molecule is then set out so the group with the lowest priority points away from you. If then you trace the groups around in order of importance, then in one case the direction is clockwise. This is called the *rectus* configuration (*R*); the other isomer when traced around will be in an anticlockwise direction and is said to have a *sinister* configuration (*S*). A simplified priority order is I, Br, Cl, F, OH, NH_2, COOH, CHO.

Orbit of electrons
It is assumed that the electrons of an atom are orbiting the central nucleus in fixed pathways similar to the planets around the sun.

Organic chemistry or organic molecules
The study of the compounds of carbon, of which there are many millions.

Osmosis and osmotic pressure
The process of solvent molecules passing through a semi-permeable membrane from the more dilute to the more concentrated solution. The osmotic pressure is the pressure needed to be put upon the system to prevent this movement of solvent molecules.

Oxidation
The addition of oxygen to an element or compound or the removal of hydrogen. It also defined as the process of electron loss. Reduction is the opposite of these statements, i.e. the loss of oxygen, gain of hydrogen and the gain of electrons. Whenever oxidation occurs then reduction also occurs and the oxidizing agent is itself reduced.

Oxide
A compound of an element with oxygen. Oxides of metals are usually basic and those that dissolve in water are called alkalis.

Oxides of nonmetals can dissolve in water to form acids; these oxides are called acidic oxides, e.g. CO_2. Some elements also form neutral oxides, e.g. hydrogen oxide, H_2O, or water. For superoxide see *SOD*.

Oxygen   A gas that comprises 20 % of the air. It is essential to life.

Peptide   A peptide is a compound formed when at least two amino acids condense or react together. A 'peptide' group is so formed, $-CO \cdot NH \cdot -$. A larger compound containing thousands of amino acids forming peptide bonds is called a protein.

Periodic table   The systematic classification of the elements based on their atomic numbers.

Peroxide   A compound containing the $-O-O-$ group. They are strong oxidizing agents. When produced accidentally in the body as a result of a carbohydrate being oxidized, the enzymes catalase and peroxidase in the blood immediately decompose it to prevent damage to body cells.

PET   The analytical procedure 'positron emission topography'. It is a noninvasive method frequently used to trace the workings of the inner body, particularly the brain.

pH   A numerical value that describes the acidity of a solution. pH 7 is 'neutral', and values 1–6 are acidic and 8–14 alkaline. $pH = -\log_{10}[H^+]$.

Physical properties   The properties to do with the melting and boiling points, solubility, etc. All the processes that are reversible.

Pi bond   Sometimes called 'double bonds', they predominantly occur between two carbon atoms or between carbon and oxygen.

Polymer   Very long chains of like units, as in 'polythene'.
Polypeptide   Chains of amino acids joined together with peptide bonds.
Polysaccharide   Chains of sugar-like compounds as in cellulose (see *monosaccharide*).

Precipitate   A fine-grained insoluble substance that is formed when two clear solutions react together to form an insoluble material that slowly falls out of solution as a residue.

Progesterone   A female hormone, see Chapter 6 for its structure.
Protein   A very-long-chain biological molecule made up of thousands of amino acids.

Proton   A positively charged particle present in the nucleus of atoms. The number of protons in the nucleus of an atom determines the atomic number of the atom.

Pure   A description of a material that contains only one type of substance.

Qualitative   A property of a material based upon its characteristics.
Quantitative   A description of a property of any material that is a measurable quantity.

Radioactive and   The spontaneous disintegration of the nucleus of some

radioactivity   elements giving off alpha, beta or gamma radiation. Many naturally occurring rocks and minerals contain radioactive isotopes.

Ratio   The proportion of one number to another, represented by a ':' sign. The ratio of 1 to 4 is shown as 1:4.

Recrystallization and crystallization   The formation of crystals from a saturated solution.

Redox reaction   A combined description of both oxidation and reduction processes.

Reduction   The opposite of oxidation.

Relative atomic mass   See *atomic mass*.

Saccharides   The general chemical name given to a wide range of compounds sometimes called carbohydrates. See also monosaccharide.

Salt   The compound sodium chloride, NaCl.

Salts   In general, compounds formed by any acid and any base reacting together.

Saturated hydrocarbon A compound that contains only single chemical bonds between the carbon atoms present. Unsaturated hydrocarbons contains double bonds between two adjacent carbon atoms, e.g. $H_2C=CH_2$. Many of these double bonds in a compound make the compound polyunsaturated. This occurs in some natural fats.

Saturated solution   A solution in which no more of the solute can dissolve at that temperature.

Single bonds or sigma bonds   Single covalent bonds made up of one electron from each of the two covalently linked atoms.

SOD   Superoxide dismutase, an enzyme that decomposes the un-needed enemy of the cell, superoxide anion ($O_2{}^-$), which is made as an unwanted by-product of cell respiration.

Soluble   The substance that dissolves in a solvent to form a solution: solute + solvent = solution.

Standard notation   A method of expressing large numbers as powers of 10, e.g. 1000 as 10^3, and very small numbers in a similar way, e.g. 0.00001 as 10^{-5}. Remember $10^0 = 1$.

Steroid   A series of compounds that are found to be particularly active in the human metabolism. They all have the common ring structure,

Stero-isomerism   The type of isomerism that depends upon the shape and arrangement of the atoms in a molecule.

Sublimation   The process of changing from a solid to a gas without first melting to a liquid, and the reverse of this process, going from a gas directly to a solid on cooling.

Superoxide   See *SOD*.

Symbols for elements   There are approximately 100 elements and each has been given an abbreviation or symbol. A list of the common ones appears in Appendix 1 along with their atomic masses. Some of the symbols are easily recognized from their British name, like carbon, C, or hydrogen, H, but some are named after their Latin names, e.g. sodium (natrium), Na, and iron (ferrum), Fe.

Synthesis and synthesizing   A process of putting together either directly or indirectly elements or compounds to form a new compound.

Temperature   A measure of the amount of kinetic energy of particles in a material. At higher temperatures the particles move more quickly.

Testosterone   A male hormone. See Chapter 6.

Thermal decomposition   The breaking up of a compound by the action of heat.

Tincture   A solution of a material in alcohol. It is an old alchemical name that seems to linger on.

Valence   An older name used to describe the bonding within a compound.

Vapour   The gaseous particles of a material.

Vitamins   A range of essential chemicals of the human body. They are alphabetically arranged in the order in which they were discovered. Some letters are missing because they were mis-named when discovered and found to be identical to others already assigned. Some are water-soluble and need to be taken in on a regular basis via a balanced diet. These include vitamins B and C. Fat-soluble ones include vitamins A, D, E and K. There is no common pattern to the structures of the vitamins. They all have different roles in the body.

Volatile   A substance that is easily evaporated into the vapour phase.

Zwitter ion   An ion of an amino acid that contains both positive and negative charges at opposite ends of the molecule.

Bibliography

Aspirin

1. T. M. Brown *et al.* Aspirin how does it know where to go? *Education in Chemistry*, March 1998, 47.
2. S. Jourdier. Miracle drug. *Chemistry in Britain*, February 1999, 33.
3. PolyAspirin: for targetted and controlled delivery. *Chemistry in Britain*, October 2000, 18.
4. C. Osborne (ed.). *Aspirin*. Royal Society of Chemistry, Cambridge, 1998.
5. Herbal alternative to aspirin. *Daily Telegraph* 20 September 2002, 25.

Vitamins

6. Vitamin C helps relieve stress. *Chemistry in Britain*, 1999, 16.
7. R. Kingston. Supplementary benefits – vitamins. *Chemistry in Britain*, July 1999, 29.
8. Vitamin C. *Chemistry in Britain*, August 2001, 15.
9. Fruity cancer cure. *Education in Chemistry*, November 2001, 164.
10. Vitamin C – a head start. *Education in Chemistry, Info Chem*, March 2002, 74, 1.
11. The F factor (folic acid). *Education in Chemistry, Info Chem*, January 2003, 79, 1.
12. Lead in the womb. *New Scientist*, 23 May 1998, 7.
13. T. P. Kee and M. A. Harrison. Osteoporosis: the enemy within. *Education in Chemistry*, January 2001, 15.
14. P. Jenkins. Taxol branches out. *Chemistry in Britain*, November, 1966, 43.
15. R. Highfield. How to stop cancer in its tracks. *Daily Telegraph*, November 27, 2002, 18.
16. Attacking cancer with a light sabre. *Chemistry in Britain*, July 1999.
17. Seek and ye shall find. *Chemistry in Britain*, July 1999, 16.
18. P. C. McGowan. Cancer chemotherapy gets heavy. *Education in Chem*, September 2001, 134.
19. S. Cotton. Combretaststin (anti cancer drug). *Education in Chemistry*, September 2001, 121.
20. D. Derbyshire. Stinkweed halts cancer cells. *Daily Telegraph*, 1 September 2002, Science correspondence.
21. R. France and D. Haddow. Biomaterials to order. *Chemistry in Britain*, June 2000, 29.
22. R. Langer. New tissues for old. *Chemistry in Britain*, June 2000, 32.

23. S. Aldridge. A landmark discovery, (penicillin). *Chemistry in Britain*, January 2000, 32.

24. V. Quirke. Howard Florey – medicine maker. *Chemistry in Britain*, October 1998, 35.

25. Chemists fight drug resistance. *Education in Chemistry, Info Chem*, March 2002, **62**, 1.

26. Superbug beater. *Chemistry in Britain*, October 1999, 18.

27. Time to attack harmful organisms. *Education in Chemistry, Info Chem*, March 2002, **62**, 2.

28. J. Mann. Medicine advances. *Chemistry*, 2000, 13.

29. D. Bailey. Plants and medicinal chemistry. *Education in Chemistry*, July 1977, 114.

30. R. Kingston. Herbal remedies. *Education in Chemistry, Info Chem*, March 2001, **68**, 2.

31. Garlic's healthy effects explained. *Chemistry in Britain*, November 1997.

32. Barking up the right tree. *Chemistry in Britain*, April 2000, 18.

33. M. Lancaster. Chemistry on the cob. *Education in Chemistry*, September 2002, 129.

34. J. Cassella *et al.* Harnessing the rainbow. *Education in Chemistry*, May 2002, 72.

35. Garlic, naturally. *Education in Chemistry, Info Chem*, July 2002, **76**, 1.

36. M. Jaspars. Drugs from the deep. *Education in Chemistry*, March 1999, 39.

37. G. Cragg and D. Newman. Nature's bounty. *Chemistry in Britain*, January 2001, 22.

38. B. Griggs. The cure all, clover. *Country Living*, May 2000, 134.

39. Hawaiian plant may help yield TB drug. *Education in Chemistry, Info Chem*, September 2001, **71**, 1.

40. P. C. McGowan. Cancer drugs. *Education in Chemistry*, September 2001, 134.

41. T. M. Brown *et al.* You mean I don't have to feel this way? Clinical depression. *Education in Chemistry*, July 2000, 99.

42. Re-educating the immune system (diabetes). *Education in Chemistry, Info Chem*, March 2001, **68**, 1.

43. A. Butler and R. Nicholson. Bring me sunshine. *Chemistry in Britain*, December 2000, 34.

44. Safe sun. *Chemistry in Britain*, July 2001, 58.

Alzheimer's disease

45. Nice approves first Alzheimer's drug. *Chemistry in Britain*, March 2001, 10.

46. K. Roberts. Alzheimer's disease:forget the past, look to the future. *Education in Chemistry, Info Chem RSC*, May 2000, **63**, 2.

Heart treatments

47. L. Gopinath. Cholesterol drug dilemma. *Chemistry in Britain*, November 1996, 38.

48. Anti-cholesterol drugs help arteries dilate, 20 July 1998; www.healthcentral.com/news

49. S. K. Scott. Chemical waves and heart attacks. *Education in Chemistry*, May 1998, 72.

50. NicOx says NO to better drugs. *Chemistry in Britain*, March 2001, 11.

51. J. Saavedra and L. Keefer. NO better pharmaceutical. *Chemistry in Britain*, July 2001, 30.

52. F. Murad. Nitric oxide and molecular signaling; http//girch2.med.uth.tmc.ed

53. L. J. Ignarro. Molecular and medicalpharmacology (regulation and modulation of NO production); http//resesrch.mednet.ucla.ed

54. J. S.samler. Nitric oxide in biology; www.hhmi.org/science/cellbio/samler.htm

55. US researchers win Nobel Prize in Medicine, October 1998; www.healthcentral.com/news

Miscellaneous drugs

56. S. Cotton. Marijuana. *Education in Chemistry*, November 2001, 145.
57. A. T. Dronsfield and P. M. Ellis. Ecstacy – science and speculation. *Education in Chemistry*, September 2001, 123.
58. S. Cotton. It's a knock out (hypnotic drugs). *Education in Chemistry*, November 1998, 145.
59. S. Cotton. Cocaine, crack and crime. *Education in Chemistry*, September 2002, 118.
60. S. Cotton. Steroid abuse. *Education in Chemistry*, May 2002, 62.
61. S. Cotton. Zyban (for treating nicotine addiction). *Education in Chemistry*, March 2002, 35.
62. S. Cotton. More speed, fewer medals. *Education in Chemistry*, July 2002, 89.
63. S. Cotton. Drug not dosh. LSD. *Education in Chemistry*, May 2000, 63.
64. Vick Inhaler. *Education in Chemistry*, July 2002, 89.
65. D. C. Billington. Drug discovery in the new millennium. *Education in Chemistry*, May 2001, 67.
66. Time to attack. *Education in Chemistry, Info Chem*, March 2000, **62**.
67. The Tagamet tale. *Education in Chemistry, Info Chem*, March 1998, **50**, 2.
68. N. Agrawal. A class act (more selective painkillers). *Chemistry in Britain*, August 2001, 31.
69. M. Gross. Know your proteins. *Education in Chemistry*, September 2001, 128.
70. The master protein. *Education in Chemistry, Info Chem*, November 2002, **78**, 2.
71. RSC celebrated DNA fingerprinting. *Education in Chemistry*, November 2002, 144.
72. A. T. Dronsfield *et al*. Halothane – the first designer anaesthetic. *Education in Chemistry*, September 2002, 131.
73. P. D. Darbre. Oestrogens in the environment. *Education in Chemistry*, September 2002, 124.
74. Z. Guo *et al*. Metals in the brain. *Education in Chemistry*, May 2002, 68.
75. B. Austen and M. Manca. Proteins on the brain. *Chemistry in Britain*, January 2000, 28.
76. Drugs on the brain. *Chemistry in Britain*, May 2000, 18.
77. K. Roberts. All in the mind. *Education in Chemistry, Info Chem*, January 2001, **67**, 2.
78. W. Gelletly. Radioactive ion beams. *Education in Chemistry*, January 2003, 13.
79. Marie Curie and the centennial elements. *Education in Chemistry*, November 1998, 151.
80. N. Mather. A time to attack. *Education in Chemistry, Info Chem*, March 2000, **62**, 2–3.

Further reading

1. A. R. Butler and R. Nicholson. *Life, Death and Nitric Oxide*. Royal Society of Chemistry, Cambridge, 2003.
2. A. V. Jones, M. Clement, A. Higton and E. Golding. *Access to Chemistry*. Royal Society of Chemistry, Cambridge, 1999.
3. R. Lewis and W. Evans. *Chemistry*. Macmillan, Basingstoke, 1997.
4. *Practical Chemistry*, 13 video clips of practical techniques in chemistry. Royal Society of Chemistry, 2000; available from Dr Rest, Department of Chemistry, University of Southampton SO17 1BJ (covers most of the techniques outlined in Chapter 1 of this book).
5. *Spectroscopy for Schools and Colleges* (CD-ROM). Royal Society of Chemistry, Cambridge 2000 (covers in more detail all the techniques listed in the Analytical section of this book).

Index

Chemistry: An Introduction for Medical and Health Sciences, A. Jones
© 2005 John Wiley & Sons, Ltd